Triumph in Crisis
A Medical Narrative

VICTORIA HUGGINS PEURIFOY

Introduction by Valerie Fox

Triumph in Crisis

Endorsements:

What does resilience look like? *Triumph in* Crisis describes it for us. Overcoming and overcoming and overcoming yet again, Victoria keeps her sense of humor and her faith. Her story inspires me to cope with and rebound from any medical obstacles that I may face in my journey through life.

- Piccola Etchison, Retired Maryland public school teacher and volunteer reading tutor

%%%

I was very impressed by the intimacy and insight of your fertility story. I thought that the medical details were accurate and descriptive, and the up's and down's of this journey were compelling

-Dr. William King, Pediatric Medicine

%%%

Victoria Huggins Peurifoy wears many hats; daughter, mother, wife now widow, voice-over artist, singer, poet, author and ghost writer. However, there was one hat that she never wanted to wear; patient.

In her first book, A Blade of Grace, she opened herself up and described some parts of her life so vividly it was hard not to relate to a few of the "been there, done that" moments. I for one, found myself laughing, crying, cheering and groaning as I peeked into the seams of the tapestry of her experiences.

This book, Triumph in Crisis, is not just an appropriate title; it is a look into the author's world of medical turmoil and her victories where she gives all glory to God and rightfully so. Victoria not only shares the medical challenges she faced, she also takes us by the hand and shows us how the medical personnel does and sometimes does not demonstrate compassion for the patient. She explains some medical procedures and shares why she had concerns about them. She paints a picture of frustration when she could not get satisfactory answers and shows us how to not take the defeatist attitude in spite of obstacles.

Many should read this book, regular lay persons and medical professionals. Maybe then, they will stop spelling at us and explain procedures more clearly. After reading this book, I intend to ask even more questions when I go to the doctor.

-RuNett Nia Ebo, Poet of Purpose, Author, Playwright

Acknowledgments

Rachel Wenrick was my narration professor at the Dornsife, of Drexel University. I learned so much in her class. She was sooo pushy…but in the greatest way. I fussed about writing these narratives and she ignored my complaints and insisted that I go forward. I love how she encouraged me. Thank you Rachel; for your persistence and appreciation of my efforts.

Valerie Fox, of the Writer's Room at Drexel, coached, advised, recommended, taught, and encouraged me to make this narrative the best it could be. Her constructive critiques were especially helpful. She wanted my book to be filled with facts, poetry, and the full aspect of a non-fiction narrative. Thank you for believing in me.

RuNett Ebo Gray, my Poetic Partner in Rhyme has always been my moral supporter, personal Grammarly®, book title guru, and first to tell me when something doesn't make sense. This story makes sense to her especially since she was a part of one of the narratives. Thank you for everything partner. (Author of Lord, Why did you make me Black © 1994)

Darlene S. Huggins This lady is my sister, my rock, my undeclared cheerleader. She helped with part of the title for this book and explained that I have been *Triumphant* through each medical challenge. Thank you sister…I love you. *(Recent graduate of Chestnut Hill College)*

Naomi Strange added "Crisis" to the title of this book. She has read every book I ever wrote and every article written about me. From my literary beginnings, Momma Saun, as I affectionately call her, also knows of each *Crisis* I have endured. I thank you for your constant belief in me and for your encouragement *(Retired English teacher at Girl High School of Philadelphia)*

Drexel's Dornsife – Key Spot- This is the computer room where two special individuals assist students with their projects as they use the computers. **Brenda Lewis** and **Kevin Williams** should be recognized and acknowledged for the patience they exhibit and knowledge they share with all of the students who visit this home away from dorms, classes and home. Thank you for all of your help.

Doctors **William King** and **Daphne Golding** graciously volunteered to read the section of this narrative that related to their field of medicine…pediatrics and rehabilitation. Your input was invaluable to me. Taking away from your own personal schedules to read portions of my manuscript was a blessing to me. Thank you so much.

Introduction to *Triumph in Crisis*

Victoria Huggins Peurifoy leads a life imbued with the spirits of poetry and family. In *Triumph in Crisis: a Medical Narrative*, she shares with us insights she has gained from facing medical challenges. These challenges, albeit unintended, add wisdom to her story of life. Peurifoy brings her poet's love of language, and of the particular, to her subject. The arc of this journey is always present, yet we notice themes and imagery crossing and intersecting in memorable ways. She includes retrospective poems in addition to the core essays in order to highlight powerful images and personal growth. As a reader, you will become invested in this growth.

Peurifoy's voice blends the past, the present, the future. The past is put out in compelling detail. Her present voice of experience, comments on her stories, and compassionately so. For the future, she asks her readers to bear in mind her stories and to always treat those facing the uncertainty and anxiety of illness, health challenges, and mobility challenges with respect and patience.

For instance, she shares with us the ups and downs of fertility treatments, pregnancy, and childbirth in "Great Expectations."

We will not soon forget both the frustrations and joys she shares with us. In the delivery room, with her trademark humor, she calls her nervous husband, "Dr. Peurifoy." She quips with her actual doctor, who when giving her a pep talk, refers to "our" hard work. *"Our hard work,* don't you mean *my hard work?"* Peurifoy makes timely and serious points—including some that are critical of the medical community—so that practitioners, and indeed everyone in the larger medical community, will understand her view and take her seriously.

This work is an important addition to the expanding body of work known as medical humanities. It's easy to say: healthcare workers should be empathetic. The concept may be harder to apply. Such application requires reflection. Reading Ms. Peurifoy's *Triumph* may bring about that reflection and by extension, a question and a change.

--Valerie Fox, Teaching Professor, Drexel University and Writing Fellow at Writer's Room Dornsife.

Foreword

Rachel Wenrick

Of the many things I love about Victoria Huggins Peurifoy and her work, first among them is her voice.

As a child, she talked—a lot. So much so, that her mother gave her a pen and paper and told her to start writing. (Lucky for us, she never stopped.)

As a performance poet, she is known as the Axiom of Truth—the root of understanding.

As a singer, she lives the song. It comes through her.

This book is about being heard.

Victoria understands that those of you in the medical profession are *busy*. You have so many pressures to contend with, but your patient—this person in front of you—comes first.

She is here to hold you accountable.

By you, I mean us.

By her voice she has given us her word.

By her writing, we are moved to keep ours.

Index

List of Medical Challenges

1972 -- Breast tumors in the left breast caused a humongous scare but the results were benign.

1974 – Left breast expanded to the size of a balloon, benign cysts

1981 – Bones in my feet corrected

1983 – C Section   twin daughters born. (One natural and <u>one</u> by C –Section)

1987 – C Section of my son, who had lain transverse across my stomach.

1994 -- Cervical Laminectomy to remove part of my vertebrae, which allowed my spinal cord room without compressions.

2001 -- Brain Surgery – 60% of vision lost

2008 -- Right knee replaced

2010 – Pacemaker – I kept blacking out… heart rhythm kept slowing down

2010 - (August) left knee replaced

2014 – Carpel Tunnel of the right hand

2015 -- Trigger finger of the middle finger of the left hand

A Thoughtful Prayer...of Sorts

As the index indicates, you can see, I have had some medical challenges in my life. However, you will only read six narratives in "Triumph in Crisis" which dramatizes the life this author has experienced as a patient. This narrative chronicles major medical situations that catapulted me into surgeries.

Patients, in general, often have many problems long before they actually agree to any surgery or dramatic procedure; designed to make their big boo-boos all better. Once a patient agrees to these healing opportunities, sometimes… afterwards…things happen that the patient wasn't told could or would happen.

My prayer is that nurses, doctors, interns, physician's assistants and nurse practitioners, who read these narratives, also as a part of sensitivity training, will begin to consider how their patient may feel about things to do and not do, say and not say to patients. In addition, I need to tell the whole story from my vantage point of having been the patient and how it felt to sometimes not be heard…when I should have been heard…I know my body.

The main life instrument that was my saving grace was my faith in God. Each surgery was preceded with a prayer over the hands of the doctors and their entire staff associated with healing me.

Doctors and their staff need prayer too, whether they believe or not.

Many medical practices are refined today, but knowing the history of former practices has allowed for many new interventions…and that's all good. The current mayor of Philadelphia, Jim Kenney says,

"You are never really happy unless you're helping other People."

If this book is able to help others, I will be jumping for joy. Through it all, God has been good to me. I pray that not only does the reader gain knowledge; the hope is also that you enjoy the candor and the voice of me…the patient. Amen

This page left blank on purpose

I.   Cyst Brigade

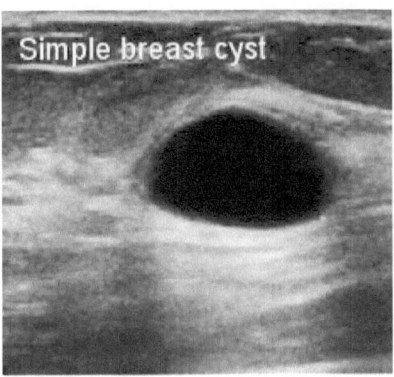

"You may not control all the events that happen to you, but you can decide not to be reduced by them."

Maya Angelou

This page left blank on purpose

Cyst Brigade

As I lay on the table in my gynecologist's office in the winter of 1972, she had already done a pap smear and had proceeded to examine my breast… first the left, then the right. However, when she examined the left breast, she frowned as though something was stinking, and then she said,

"Put your hand here." Here…was on my left breast. "Do you feel that?" She asked.

"Yes, I do. What is that?"

"You have a mass in your breast. I'm going to have to do an ultrasound to see what's going on.

I had one of those on-general-principle conversations with my mother about my paternal grandmother and how she suffered with cysts in her breast and was always having operations. However, mom didn't speak about breast examinations. Breast cancer was not evident in our family, but she surmised that I needed to know the medical history of my father's family.

The doctor asked, "Do you ever examine your breast?"

"No. I never have," was my response.

She took a device and put some kind of gel-like substance on it. She rubbed it across the area where the mass lay. As she was doing that, I looked at the screen and all I could see were three solid black masses that looked like circles. My doctor said these little circles looked like cysts. She explained that she didn't normally see cysts in women my age. Most times, the women are in their 40's. At the time, I was only 20 years old. I explained to her what my mother had told me about my grandmother.

The doctor said, "That's all the more reason why we need to look at these a little closer."

I wanted to know if what she felt was cancerous and what she could tell from looking at the ultrasound screen. She explained that a cyst is smooth and somewhat pliable; in other words, it would move around a little bit under the manipulation of her fingers.

She also said, "Cysts are usually non-cancerous fluid filled sacs, but we need to study the fluid for a more accurate result."

I followed my doctor's recommendation to see a doctor at Hahnemann hospital on Broad Street in Philadelphia. This visit proved to be a little bit unbearable.

The doctor on staff at Hahnemann said, "Victoria, we are going to have to aspirate the area where the mass is."

 I didn't really know what that meant so I said,

"Speak in English aka put what you're saying in laymen's terms."

The doctor continued, "What that really means is, we will insert a needle in the affected breast and suction out the fluid within to test it."

The very thought of what he proposed was frightening.

"Will it hurt?" I asked. The doctor simply stated,

"There will be some discomfort."

I interpreted that to mean, *it's going to hurt like hell!*

Discomfort

He lied! I was super sensitive to pain and that needle hurt me like the 100 fold sting of a bee. To me, the needle was big-- a 21 surgical gauge to be exact. I didn't like needles anyway. As he pulled back on the needle, fluid came out and it was bloody. At that point the doctor began talking to me about having surgery and how they needed to do a biopsy to see whether the cysts were benign or cancerous. He told me I would get a call from his secretary about scheduling the surgery, to look at the mass closer under a microscope.

A Conversation with Mom

I spoke to my mother about the surgery and how scared I was.

"Mom, what if it's cancerous, my God, what am I going to do?"

My mother tried to console me by saying,

"Your grandmother was never diagnosed with cancer after they examined her cysts. She would have the procedure, they would remove the cysts, check them. They would let her know whether they were cancerous, and they were always benign. Don't get yourself all tied up in a knot worrying about this. Have the operation and see what's going on and I'm pretty sure that everything will be fine...you'll be fine."

I loved her confidence about the situation; but I was still anxious and unsure. My thoughts were: *what if it is cancerous and I lose my breast or what if I have a big indentation where they removed the cyst. If they are benign what would I have to do after that... after the surgery?* I was left pondering.

Sick Leave

At the time, I was working as a clerk typist for the government. I had not had my job that long...only since May of 1972. I spoke to my manager about my situation. She advised me to handle my business. She promised to gather some information, to see what else would be available to help me with sick leave while I was out taking care of this. I had some sick leave but I didn't have enough to cover possibly a two to three week time off the job.

My boss said, "The last thing you need to be worrying about is this job. What you need to be concerned with is your health. That's more important. Without that, this job won't matter anyway...will it?"

The day of the surgery, I was prepped and made ready to go under the knife. The doctor had said as long as I did everything that I was supposed to do after the surgery, I would be okay. I didn't know anyone else who had this surgery...or who I could talk to, so I was in a total, internal panic mode.

I awoke from the surgery, bound by ace bandages, gauze pads, with a bag of fluid hanging overhead on a rod, next to my bed. There was no one around me. There were just a lot of bright lights in the immediate vicinity of my bed. I was in recovery, but mentally, I was not recovering. My mind said, *you have cancer and they have taken your breast.* I immediately began screaming and crying and crying and screaming and pointing toward my breast... that I just knew was gone. I was trying to feel the area but all I could feel were the pads. A nurse, who was nearby, came over to me and asked,

"What's the matter?" I didn't say a word; I just pointing and screamed.

She looked at me and said, "Stop screaming! With a hump that big, they didn't take your breast. They just have you heavily padded that's all".

Through my tears, I mustered a smile and so did she.

"The doctor," she said, "will be in to talk to you in a few minutes; so just calm yourself down. Okay?" I simply nodded my head. I was in the hospital four days.

"Ms. Huggins, we removed three non-cancerous cysts from your breast. I'm happy about that; as I'm sure you are too…right?" he asked. Then he continued, "What I will need you to do is change the pads twice a day, take your medicine as scheduled, take it easy, and DO NOT go back to work for two weeks."

I followed instructions and did not have any more problems until…1974.

Part II- 1974

Like raging lava rolling down the mountain, I was awakened by heat and pain in my left breast not realizing what my mirror would reveal. To you I will just say, imagine birthday balloons, all evenly shaped, except one special balloon is much bigger than all the others and it has "Surprise" inscribed on its surface. Without plastic surgery, my left breast went from a 36 C bra cup to a 55 DD bra cup...there was one big problem... I was extremely lopsided.

On a spring morning in 1974, one week before I was scheduled to go on the Northern Concert Tour with Community College of Philadelphia, I knew something was terribly wrong. My left breast had blown up like someone had pumped helium into it. Touching it hurt, and it was inflamed to a bright red. My first response was to panic! But instead, I called my mother.

> "Mom, I don't know what's wrong with me. My one breast is 10 sizes bigger than the other. It hurts, it's hot, and it's red. I can't get on a bra, or a top. I found an oversized t-shirt that barely fits. I don't even want the fabric to touch my skin. Mom, I'm only 22 years old, why am I having all these problems with my breast?" I asked.

My mother did not have an explanation for me.

> She just said, "Calm down, get yourself together. I will come to your apartment and take you to the doctor."

Mom Sees for Herself

When my mother saw me, she became alarmed by the size of my breast.

Mom said, "I have never seen anything like this in my life. Did you cut yourself or pierce your skin?"

"No mom. I told you, I woke up like this. Nothing bit me; I believe I would have felt it!"
 I said.

Doctor's Visit

As the doctors examined my breast, I squirmed and winced at each touch. I was so inflamed that the x-ray machine, may have only seen an inferno!

The doctor said, "You have an infection. It could be more going on but we won't be able to determine that until the infection is gone. At that point, we will x-ray you again. In the meantime, I will write a prescription for an antibiotic to kill the infection. Place a cold pack on a towel. Then lay it on your breasts. Do that three times a day…along with taking your medicine."

The Weeks Ahead

I took the medicine and used the ice pack compressions as instructed. I was already on a two week vacation; so initially, I didn't have to lose time from work. By the time my choir was scheduled to go on tour, my breast had deflated about half a percent. To wear my clothes, I had to lightly bind myself, because I still couldn't wear a bra. The pain had dissipated. I was just uncomfortable. The female choir members that I had told knew what was going on with me so they tended to shield me from being accidentally bumped. I took my medicine and continued the doctor's directions of the ice compression regiment. By the time I returned home from the tour, a week later, my breast had deflated to almost its normal size. I could finally wear my bra, blouses, dresses, and sports jackets without any problem.

The Follow-Up

When my doctor saw me, he was happy my breast size had decreased. He ordered another x-ray which revealed three cysts in exactly the same place as they were two years before.

The doctor said, "Although I am happy the inflammation (swelling) is gone, the fact that you have these cysts concerns me. We will aspirate, as you told me was done the last time. This will help us determine whether we need to do surgery."

There was blood again in the aspirated fluid, just like last time. My doctor revealed what I knew was coming.

"Victoria, I'm sorry we will have to do a biopsy to determine whether these cysts are cancerous."

A Special Occasion

My youngest sister was a senior at West Philadelphia High School. This was an especially challenging year; but, she was determined that she was going to graduate and have good grades. She informed me of the date of her graduation... June 20, 1972. I was so excited for her.

I even said to her, "This is so exciting. I can't believe my baby sister is graduating from high school. I will be the loudest one in the audience rooting you on... I can't wait."

The doctor was examining me again and said, "Thought you'd be happy to know, I have scheduled your surgery. It will be for June 19th. You will receive a call from..."

I interrupted him,

"No, no, no! It can't be that date. My baby sister's graduation is the next day... I have to be there. Can't you re-schedule it a week before or after that date? It can't be that date!" I exclaimed!

The doctor was not budging.

"I'm sorry, that is the date it is, and it can't be changed. We can't wait; we want to determine, as soon as possible, what's going on within your breast. We want to make sure that it is not cancerous. If we can catch it early and remove the breast...it'll be safer for you." he said.

At that moment, I became frantic and said.

"I don't agree to this, I want to know if it's cancerous. You have to wake me up. Don't remove my breast without discussing it with me first... I have a right to know what's happening." I cried.

The doctor said, "If these cysts are cancerous, we will not be waking you. We will already be in the position to remove the cancerous cysts. We don't want to subject you to more infection... I'm sorry... This is the reality!"

Frustration Overdrive

I went home so upset that I couldn't think straight, but the recesses of my racing mind had shifted into overdrive… *Remove the cyst, this might be cancerous, possible breast removal, missing my baby sister's graduation… it's not fair God, why me? This is all too much.*

I did not have anyone to turn to who could empathize with my fears. My mother tried to console me as did my sisters, but it was all in vain. I knew I had to deal with the situation, but my options were ugly and out of my control. I believe my thoughts, frustrations, and fears were exacerbated by *We will do what has to be done... Regardless of what you want or how you feel. That's what I heard...*even though that's not what was said.

Day of the Operation

My mother drove me to the now-defunct Osteopathic Hospital, on City Line Avenue in Philadelphia...now a school. The normal preparation took place before surgery, but my nerves were enthralled in emotions. I wasn't looking at any of this positively. My thoughts were...*am I going to lose my breast? If I lose my breast, will I be lopsided and deformed. My boyfriend won't love me anymore and won't want to touch me. I might even die on the table. I could not hear anything the doctor was saying to me.* The oxygen mask was placed on my face and I went to sleep dreaming of being naked; running through a field of lavender with the remaining breast flapping in the breeze.

From a distance it seemed I could hear someone say,

"Victoria, Victoria, wake up. You had non-cancerous cysts; that we call a fibro adenoma of the breast. We did not have to remove your breast. However, we did bind you to insure against excessive bleeding."

I think it was my doctor speaking to me, but all I heard was...*non-cancerous...We did not remove your breasts.* "Thank you Jesus, Lord, God." I prayed a prayer of thanksgiving.

After Surgery

On the second day of my stay at the hospital, my pain medicine had just kicked in, when my baby sister waltzed into the room on cloud nine with her graduation cap and gown. In her hands were a beautiful dozen red roses that accented her glow. My mother and other sister Dar came in behind her. As my youngest sister came toward me grinning, She said,.

"Vicky, here's my diploma." She smiled, and handed me the roses.

"These are for you. How are you feeling?" She asked.

After I congratulated her and told her how proud I was of her, I told my family how petrified I was that I was going to lose my breast. I told them that I could now breathe a sigh of relief.

My mother looked at me and smiled and said, "You did all that worrying for nothing... and here you are perfectly fine."

She may have been right, but she wasn't me. I wondered if she would have reacted any differently if all of this had happed to her.

Post-Surgery

Post-surgery, my doctor shared with me that when I originally presented my oversized breast, he was sure I had cancer. He never mentioned it because he didn't want to scare me or jump to conclusions.

I said, "You didn't do a good job of that because I was already petrified."

Further he said, "You will need to keep the surgical area clean as you've been doing. Your dressing will still need to be changed three times a day or more as necessary. If you have any worse pain, any redness, swelling or drainage from the site, go to emergency. You can return to work in a week, but no heavy lifting of any kind; no jogging or rough activities for two weeks."

By the end of six weeks, the incision had flattened and there were no indications that I would have any other problems.

Forty-one years have passed since both incidents with my breasts; I have not had any more scares. Frequently, I speak to young women about being aware of changes in their breast. I also share how important it is to see a doctor right away if, when checking themselves, they see any changes of any kind in their breasts. Remember, vanity is not as important as your health and life. I can calmly say that now.

Fear of Cancer

Two breast cancer scares

Electrocuted my heart's fear.

First cyst appeared from nowhere,

Lumps caught during a gynecological exam.

Until you hear a biopsy is needed

You are perfectly calm. But say those

Words and you are propelled into a

Well-oiled imagination.

Everything turns out alright, then, you think

You're safe.

Two years later, you

Awake to one breast, that is so

Enormous, you fear movement will

Make it break. But, it is so hot, so

Painful to the touch, you sense your

Doom. Cancer is your only surmise.

Antibiotics, biopsies and more,

You go under the knife…once again.

You argue with the doctor

To wake you if it is cancer…

But don't take your breast.

He says…

"No, we will take it, and

Dare not wake you for fear of infection."

"Don't I have a say?" were my words?

Apparently not!

The fear, anxiety, and tears were all for

Naught.

There was no cancer

I kept my breast!

© February 1, 2017

II. Great Expectations

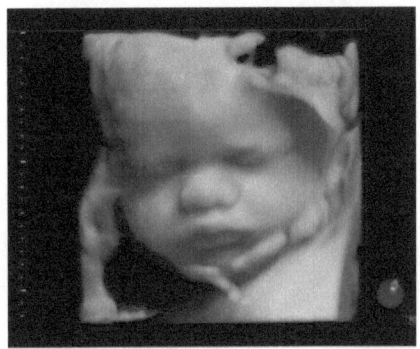

"The future depends on what we do in the present."

Mahatma Gandhi

This page is blank on purpose

Great Expectations

My husband and I had been married for two years and hadn't done a thing to stop us from conceiving. Yet, we seemed to have run into a snag. Because, alas... we hadn't gotten pregnant! So I went to a specialist in the field of infertility.

Let me take you back to a moment in 1969. This was the year I graduated from high school. It was also the year that I found myself pregnant with my first child, a boy. I was scheduled to deliver my baby two months before my graduation. This was a baby that I would never get to know, get to touch, get to hold, get to hear him coo or anything a mother gets to experience. The reason...he was given up for adoption by a child (me) who could not take care of him.

When I got married, I desperately wanted a child. Fortunately, I married a man who also loved and wanted children. Both of us, my husband and I were older than most of our friends when we got married. He was in his mid-30s and I was in my late 20s. Having a baby was our first order of business in our married life.

Our first visit with Dr. Wood, of Thomas Jefferson Hospital, in May of 1982, proved to be quite interesting. To prepare for my visit, I did something that I felt would be useful to the doctor.

I had a very erratic menstrual cycle. It came whenever it remembered! I nicknamed my menstrual cycle "Aunt Rose."
I took the time to go back to my calendar for the years 1980 and 1981 and up until April of 1982. I charted the dates that Aunt Rose had visited. It revealed that she paid a visit between seven or eight times a year.

Nevertheless, at our first visit, Dr. Wood asked us many questions, including a very peculiar one.
     "Do you know how to get pregnant?"
My husband and I both looked at each other and said,
     "Yes! That was a strange question."
     The doctor said, "Actually, it's not as strange a question as it sounds."
The doctor was speaking technically from the standpoint that in attempting to get pregnant, I should have been counting 14 days from the first day that Aunt Rose began, and then from the 14th to the 17th day, I was supposed to be able to have mad fun with my husband; then like magic, I was supposed to be pregnant. However, there was no baby bun baking in the oven.

Anyway, I said, "Yes. I do know how to get pregnant. However, nothing is happening."

I decided it was time to give Dr. Wood my little chart. He looked over it and said,

"No wonder you haven't gotten pregnant, your menstrual cycle is all out of whack."

He proceeded to tell me what he was going to do. From that point on, I kept a diary of everything I went through to conceive. What follows is based on that diary.

%%%

5/13/1982

The nurse took five tubes of blood to perform various tests. The doctor performed an internal exam and a Pap test. He explained that an M.R.I. would be administered… which was an x-ray of the frontal lobe area of my brain. Further, after all test results were received, he would prescribe medication to make the eggs that dropped during ovulation, better for fertilization. My last cycle was 4/24/1982, so the doctor wanted to see me in two weeks. The waiting would give me time to think about my future and getting pregnant. Also, at this appointment, the doctor discussed the "M*ale hormone*" factor in all women; which is located in the frontal lobe of the brain. He explained that if this hormone factor was higher than normal, it could affect my getting pregnant. He also gave me a basal temperature chart that he wanted me to use to record my temperature each morning. I wasn't supposed to put my feet on the floor before taking my temperature. When I opened my eyes…in my mouth… went a thermometer.

Basal Temperature Chart

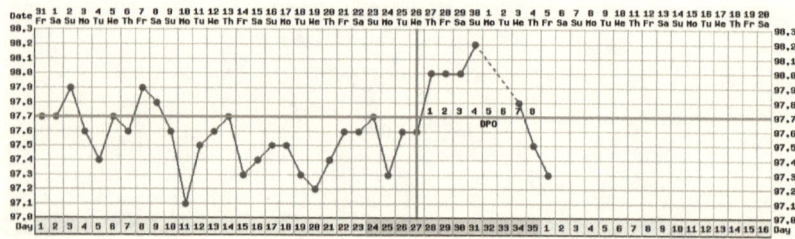

When I returned for my next visit on May 27th, the doctor found that Aunt Rose still had not paid a visit. So, to stimulate my cycle, he gave me an injection in the hip to try and force my menstrual cycle to start. It was supposed to appear in about five days; after which time, I wasn't to take the special medication. I understood that if Aunt Rose didn't come, I was supposed to take the medication and I was to call the doctor. At this visit, the MRI of my brain was performed.

June 3, 2002

I spoke with Dr. Wood about Aunt Rose still not visiting. He indicated that he had received the test from the M.R.I. and he didn't like the results. He stated that on one level, the results were normal. However, the hormone levels were higher than normal. He began talking about an experimental drug, that was designed specifically for women like myself who had the problem that the test indicated. Although he wanted to discuss it more at our next appointment, he did mention that the drug was $1,000 a month. I thought that was crazy! *"Women like myself..."* All I could think was, *I must be a really weird person or something. But we'll see.*

June 10, 1982

Dr. Wood had a lot to tell me and I had a lot of questions. Three major things were discussed:

1. the results of the M.R.I. that revealed excessive Male hormones
2. a prescription for better eggs
3. the experimental drug

Many terms were so technical, it was all overwhelming. The doctor wanted to explain in depth what the x-ray revealed and that my body was producing too high a quantity of male hormones. This could be the cause by several things...

- The reason Aunt Rose came to town whenever she got ready, and
- Stress from the job or personal problems or any other applicable situation.

My husband and I were having a grand time. I couldn't think of anything that would be causing me any stress... So here I go thinking again.... *Hypnotism would be next...just thinking out loud.*

The doctor said the X-ray had shown un-levelness (imbalance) of fluid in the frontal lobe of my brain, and recommended a repeat of the same test to see if the results would be the same. Now, I figured they're trying to say I'm unbalanced and everyone knows that's not true...although, some of my friends might agree with the test results.

Next, the doctor prescribed this medicine called dexamethasone. This was a very potent drug which was to help produce more of what I needed in the way of female hormones.

Regarding the dexamethasone, he warned, "This medication has a tendency to make you very alert."

I expressed concerned about not being able to sleep and he assured me that I would not have a problem with sleeping. I was to begin taking this medication on the 25th of June. The medicine was also supposed to make my eggs better.

I questioned, "Can I become pregnant with this medicine?"

His response was, "It is possible, but not probable."
I thought to myself so *much for encouragement.* Truthfully, I was looking for an easy way out.

In addition, the experimental drug, which could be induced only by way of an injection, was geared specifically to women whose male hormone is too high.

Dr. Wood explained, "The male hormone chemically looks like "FSH" and the female looks like "LH."
In my case, I had too much FSH. The experimental drug would give me more of the LH that I needed.

So I asked, "Isn't there a drug on the market already; that I could use."

He explained, "There is a drug on the market however; it would not be as helpful because, it has both the FSH and LH. You already have too much FSH. Taking another drug would not produce the desired results."
I was concerned with side effects of the experimental drug. The doctor said, every aspect of taking the drug would be explained. The experimental drug:

1. It had the LH that I needed. As a matter of fact, that was the main ingredient.

2. Taking the drug would help with the FSH factor... and it was safe

3. Close observation would be made of me; which included blood tests, injections, and ultrasounds. Incidentally, I had ultrasounds before; however, I didn't like idea of injections. I expressed this feeling to the doctor and he explained that...

4. The ultrasound was necessary to view the ovaries. The LH injection could cause the ovaries to become enlarged; if the patient was not closely observed...This was the side effect. The doctor required that I commit to at least six months of close observation. Therefore, he wanted me to start in September if I decided I wanted to go through with these injections.

He made every effort to have me understand that if I became pregnant while I was using this drug, the medication would then be stopped. The reason was that we would have achieved our goal. In addition, if I agreed to this experimental drug, there would be no cost for anything to us... that was the best news yet.

The initial drugs that were going to be use were:

- Dexamethasone – is used for Ovulation Induction. None ovulation is one of the more common causes of **infertility**. Clomiphene citrate (Clomid) is a medication that helps with ovulation. This medicine was to be taken with milk and I was also to watch my ankles and face for swelling.

- Prednisone - this was a steroid. I would have to watch my weight while taking this medication. I also had to avoid all salt intake. I had to watch out for my mental state as well, because this drug could cause depression

I was beginning to ponder if having a baby was worth all of this. I had gotten pregnant so easily the first time; why all of this now that I'm married and can have as many babies as my husband and I want.

After this visit, I had time to think about the next questions that I wanted to ask the doctor: Regarding bringing up the levels of the LH by way of the experimental drug, what happens when I do get pregnant? If I have too much FSH in my system and the drug at the onset of pregnancy, do I still need the FSH and LH in my system during the pregnancy to carry the baby?

I asked the doctor, "How do you ensure that the chemical levels are accurate?"

The answer to my second question was "No," because the pregnancy will decrease both of the chemicals and the body will begin to create new hormones for the baby.

Regarding the basal temperature chart, signs of pregnancy would be evident in the persistence of elevated temperatures beyond a two weeks period, as well as a missed cycle. Those are early indications of pregnancy.

6-25-1982

This appointment was set for me to have more X-rays performed. The last time I took these X-rays, I was sitting up. This time I was lying down so they could get a better shot and hopefully better results.

Note: As I was going through this diary, that is over 30 years old, I realize I actually had forgotten many of the facts. Journaling is so important.

7-8-1982

The medication was stopped to see how my ovulation developed until July 27, 1982. After a long discussion with my husband, I decided to try the experimental medication. The doctor would need to see me two times per week, twice a month. He explained that if my ovaries were too enlarge, it could cause multiple births which is not what we wanted. So he explained that observation by him would be very important.

7-29-1982

The doctor received the second set of M.R.I results of my brain. It revealed normal variations, of what, I don't know, but it made Doctor Wood very happy. Unfortunately, the blood test taken on July 8, 1982, revealed that I was not ovulating. At that point, I was very frustrated. The doctor didn't want me to worry. He wrote another prescription for the dexamethasone. Only this time, I only had to take the pill once a day rather than four times a day.

In addition, the doctor had another blood test completed that day, to do a regular check of my body functions to ensure that everything else was working normally. He wanted to make sure that my lungs were clear and he wanted to see if I was anemic. At my August 11th appointment, he planned to check my fallopian tubes. If my tubes are not open, then doctor would not be able to put me on the experimental medication in September like he wanted to do. At the same visit, the doctor gave me a container and a form to have my husband to give a sample of his semen. I didn't know how well this was going to go over with my husband.

8/11/1982

I didn't keep this appointment because Aunt Rose came to call.

Dr. Wood rescheduled the appointment for August 23, 1982.

When?

Beginning to feel like a failure…

No baby in the belly to shout

To the world about. So many pills, x-rays, injections,

Blood test to take. Seems the only

Thing I do well these days.

I just wish people

Would stop asking;

"When are you going to get pregnant?

Don't you want children?

What are you waiting for?

The next person

That asks that question,

Might get punched out.

©February 14, 2017

My husband thought it was amusing to be asked for his sperm,

but when I frowned at him, he said,

"Victoria, I'm just kidding. Are they going to give me some nasty books to look at.?" I hit him across his arm!

8/23/1982

On this day, I had to take what I considered the most painful x-ray ever. They shot dye into me; so that they could watch the dye go through my uterus and my Fallopian tubes. On a small Sony television screen, I watched the dye filling up inside of me. The dye never went into my tubes like it was supposed to. When Dr. Wood showed me the finished x-ray, he explained that there was some type of blockage within the entrance of the tube, I immediately became unraveled. This blockage, he said, was either spasmodic or it was because of some type of infection. The next X-ray that they had to do was called a laparoscopy. In addition, he said, they would have to perform a Tubal Lavage (which was the washing out of an organ and a D &C, a medical abbreviation for dilatation and curettage. I began thinking again… *I'm ready to run and hide.*

With the laparoscopy, they went through my navel to see my ovaries and everything else in that area. This test was designed to see what was blocking the fallopian tubes. The tubal lavage was the flushing of the tubes and the D&C was designed to clean the lining of the uterus. At this point, all I could do was pray that the results of this minor surgery would not result in any other surgeries.

While I was with the doctor, he shared with me that he had gotten my husband's sperm specimen and he noted that everything was normal. All I could say was...*thank God*. I thought if anything else went wrong, I would have snapped like a bridge cracked by an earthquake. The blood test that had been taken showed that I had good blood. However, the doctor said, I had a lot of blood for a woman... whatever that meant! At the time, I didn't even think to ask what it meant.

When a woman has a lot of blood...what could that mean?

In writing this narrative, this statement made by the doctor, caused me do some research. I found that blood test that show large quantities of blood might reveal elevated levels of the iron rich protein in red blood cells that carries oxygen hemoglobin. It also is referred to as Polycythemia Vera. A secondary amenorrhea is the missing of menstrual cycles.
Look at that....I found the answer

Source: Mayo Clinic diagnosis 7dxc 2030 7490 February 2017

"Dr. Wood, will the flushing prevent me from having to take the medicine? I asked.

He said, "No, it won't. Once the tubes are opened, the medication will still be needed to make you produce eggs properly... which you still aren't doing. "

My surgery was scheduled for September 21, 1982 at 6:30 a.m.

9-21-1982 Laparoscopy, Tubal Lavage, and a D&C 6:30 a.m. Though I was on time for my operation, I wasn't taken to the operating room until 2 o'clock in the afternoon. It was at that time that they put me under anesthesia. However, I don't remember anything...except they didn't tell me to count backwards from 100. Before I realized anything had happened, someone was saying,

"It's all over... all finished." I heard someone speaking as I was being wheeled back to room 468 in the old section of Thomas Jefferson Hospital's short procedure unit. I saw Dr. Wood who told me that he would call me later in the day to let me know the tests results.

9/22/1982

Around 9:30 a.m. the next day, the doctor called and explained that the dye didn't go into my tubes on the August 23rd x-ray. He indicated that it wouldn't flow through because I might have been spasmodic. He was hoping to see that the laparoscopy would show that the right tube was perfectly fine and the left tube would have no problem at the end of it. The dye only dribbled through the tube. Also at the August 23rd appointment, I was extremely tense; that's why nothing would cooperate. The blockage that was seen was thought to be caused by some type of infection; but, no infection was seen; so he did not feel the surgery was necessary.

However, he did find a surprise. My womb had a fibroid tumor the size of a golf ball, which meant that it was growing. Earlier in the year, at a prior exam, my GYN doctor performed my Pap test. That doctor told me that I had a tumor the size of a coin dollar… now it's the size of a golf ball! Again, the doctor didn't seem to be very concerned. No need for surgery based on this information. The doctor explained that we would be able to start the special medication designed just for women like me. But of course, he discussed that with me at our next visit. In addition, he also started me back on the dexamethasone. (Remember, it was supposed to help me ovulate)

My husband was very happy that I was alright and would be able to conceive. However, at the moment, I was not happy…I had

Band-Aids® all over me. I had them on my left hand and arm, covering the area where they took blood. One was on my navel to cover the incision where they were looking at my tubes and one at the hairline of my pubic area after they looked at my ovaries. *Maybe*, I thought, *I should get some stock in Band-Aids®.*

I actually was beginning to feel better knowing that no more procedures were warranted. I had even begun to pray for a baby that wasn't here. even though I had been praying from the beginning. I was telling the un-conceived baby that I loved it. I felt in my heart that conception was soon at hand.

10/7/1982

Doctor Wood had given me a diagram of the surgery results which exhibited my womb, ovaries, and fallopian tubes. He explained that the right ovary was docile, tired, sleepy, and lazy…it needed to be awakened.

(I wondered…what put them to sleep.) The research medication was called Urofolliyrophin aka FSH which was a follicle-stimulating hormone. That's the medical term. In layman's terms it is an overly emulating hormone designed to induce ovulation. The treatment would involve me receiving injections in my hip every day. I would then begin the injections, and they would be administered twice a week until the doctor advised me to stop. Everything hinged on the results of the blood test and ultrasound exams.

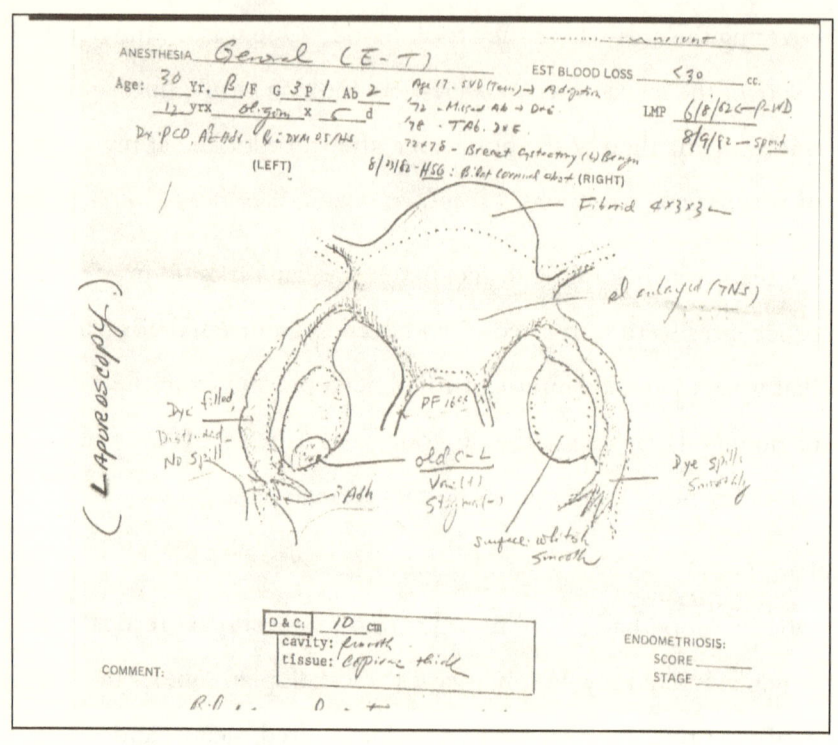

10/11/1982

Originally, I was going to learn how to administer the medication, but I decided to let my husband do it. He had experience giving his aunt insulin shots for her diabetes. My husband was going with me so he could learn how to administer the medication. I refused to give myself the shot. I was afraid. The nurses gave him instructions and allowed him to demonstrate how well he could do it. I began the injections on Monday October 11, 1982. He did a fantastic job; he didn't even hurt me, and he told me I was a good patient.

The doctor explained everything to my husband and me so that we would understand everything that was going on and what he and I should expect. I had to see Doctor Wood again and he and his assistant would be keeping their eyes on me.

10/21/1982

I had another ultra sound…what a bummer. This test was difficult at best. I had to drink six glasses of water and then hold my bladder. That's right. Hold it. I couldn't go to the restroom. I held my water from 8 o'clock a.m. until 9:45 am. Then I got angry because I had to wait so long to relieve myself. I managed to comply…which made the doctor happy.

He said, "You have ovulated, and I like the result of your temperature chart. You will not have to take the injections for a few days."

That was perfectly fine with me.

10/25/1982

My temperature was still up. That was a good start. The doctor was hoping that my baby was beginning to brew! If my cycle didn't come on by November 1, he would be elated. He planned to do more blood tests, so I had to come back on November 4th.

10/30/1982

Aunt Rose came… full blast. I swore it was trying to make up for lost time. The last time she had visited was back in August. All of November was more visits, more injections, more waiting, more hoping, more anticipation, more internal frustration, more pondering, more conjuring, more praying, more talking, more waiting. In-other-words... nothing new!

12/5/1982

Aunt Rose came again today… after I had a 98 degree plus temperature for more than 15 days. I was so upset I didn't want to talk to anyone. I didn't even want to think any more about getting pregnant.

12/6/1982

I went to see my doctor today; I was still upset about my cycle coming. We had been working at having a baby for six months. In Dr. Wood's office, I totally broke down. I think the injections were getting to me. In my mind, the whole process wasn't fast enough

Tearfully I cried, "I'm tired of this, the monitoring, the temperature taking and especially, more importantly the shots. I cannot take it anymore Dr. Wood."

My husband was trying to console me.

Dr. Wood was retorting, "No, no, no, no. You cannot work yourself up. You will ruin all of our hard work. Please don't be so upset. You have been doing a great job. And, I like the results of your temperature chart."

I fired back "*Our hard work*, don't you mean *my hard work?*"

I began to protest, "Look, I'm the one dealing with the physical aspect of this stuff. I am through with it. The Lord must be through with me too. I had a baby ten years ago and gave him away, now God is punishing me."

Tears of frustration began to flow faster than they could go over the waterfall of my heart.

The doctor and my husband felt I should be proud of myself rather than allowing me to get myself get upset by crying; because I had responded very well to the injections...so far.

He suggested, "Victoria, get all of your crying out now…then forget about it, because any stress will complicate an already difficult situation."

I retorted, "I am done. I don't want to do this anymore. It is taking too much out of me… wrecking my nerves. It is stressing me to the point of distraction and I just can't continue."

The doctor kept saying, "Don't give up now. You've been doing so well. You're going to be fine... we're doing real good right now."

I just kept saying to myself, we, *aren't doing anything...We're not getting the shots...We're not a nervous wreck...I am...forget this we stuff...I'm done... screw this!*

12/27/1982

We had a Christmas party with all of the fixings. It was an enormous success. The choir that I was singing with had come to our home with food and beverages and everyone was dressed in semi-formal attire. We had so much fun, eating, laughing, and reminiscing about past concerts and travels. By the time everything was over, I was somewhat fired up and tired at the same time. My husband had bought me a small bottle of a cordial called Brass Monkey®; which was a drink infused with rum. I told him I wanted a small amount to help me relax. But instead of relaxing me, it made me amorous!

1/6/1983

When I woke up on this particular morning I felt terrible. My stomach and my back were talking to me. Especially my stomach, it was so tender. I called the doctor to tell him of my discomfort and he wanted to see me right away.

When I got there, he had me to take an ultrasound which showed that my ovaries were swollen. Dr. Wood suggested I not do any exercise or lifting. So that's what I did all weekend. I simply tried to take it easy.

1/8/1983

Dr. Wood was saying, "Mrs. Peurifoy, we did it."

"We did what?" I asked.

"You are pregnant."

My head went down because I didn't believe him.

Dr. Wood asked, "Why so down? You should be happy. We did it, we did it, be happy." He told me.

I was afraid to be happy because I had been told once before, not long after I was married, that I was pregnant, but… I wasn't. I did not want to be disappointed again.

When I arrived at home and told my husband we were pregnant, he began hollering and screaming with delight. He ran through the house and proceeded to go outside. Out there, he jumped off the top step to the bottom step yelling,

"We're pregnant everybody, we're having a baby!"
I simply peeped out the window and went back into the kitchen to sit down. He finally joined me.

My husband came in smiling and said, "Thank you Victoria. You have made me the world's happiest man." I just smiled.

I began to feel like a baby might really be brewing, because my breasts felt tender and sometimes when I was eating, I began feeling achy in my throat and I would lose my appetite. Plus, on two occasions, I had crazy cravings. First, I wanted shrimp fried rice and then I wanted pizza. Just thinking about it made me feel fat. That's one thing I did not want to do when I did get pregnant. I didn't want to become as big as a blimp.

Now that we knew for sure that we were pregnant, we began talking about names for the babies. If it was a boy his name would be Gatewood Phillip, if it were a girl, her name would be Tory or Nicole. We chose names that were from our respective families.

1/9/1983

I received a phone call from Dr. Wood saying,

> "Well, its definite…we will be looking to see your monthly cycle in 10 months. You are very much pregnant. You should expect your baby on or about September 17, 1983. I will be keeping an even closer eye on you for the next

three months. You will only have to come into the office once a week."

He would be taking more blood tests to see if we were having more than one baby. Personally, I would have been fine with one at a time. However my husband would love to see two! But then, there was more…

"Mrs. Peurifoy, I have been looking at your test results and your blood levels are too high. They should be around 150 but instead, they are 300." Dr. Wood said.

Instantly concerned, I asked,

"What does that mean?"

"What it means is that you may be having more than one baby."

I could not even respond to him as he continued…

"I hope it is not more than two, but I have to warn you, you can be carrying up to four babies! "

"Four babies…My God…My God…say it ain't so!"

1/13/1983

As I lay on the table, all jelled up, I was watching the monitor…not sure what I was seeing. The technician conducting the test was a very handsome Italian gentleman, who sounded as if he had just come to America: he had a strong

accent. Finally I asked the olive skinned, dark haired technician,

"What's that?"

What I saw on the screen were two dots. As calmly as if he

were telling me the time of day, in his strong, just off the ship

accent, He said, "Those are your babies."

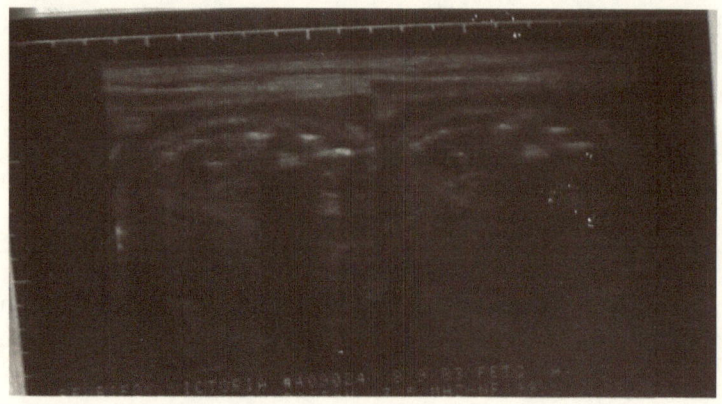

In my mind, as if a bell were repeatedly ringing, I heard, *Babies. Babies! Lord have mercy…My husband's wish has come true.*

In retrospect, I was reminded of a conversation that took place during our courtship.

One day my husband said, "You are made of good stock."

I retorted, "What am I, a piece of beef?"

He clarified, "No…that's not what I meant. What I'm saying is, you have good bone structure and you would carry twins quite nicely."

"Don't hold your breath buddy." I said.

How could I have known what God had planned?

When I arrived home that evening, I didn't mention that I had spoken to the doctor. However, my husband knew something was up because the doctor had called the house for me and had left a message. My husband inquired,

"What did the doctor say? How are you doing?"

I wondered out loud, "Why are you asking about the doctor; I didn't have an appointment today?"

He said, "I know you didn't, but he left a message on the phone for you to call. He said he would try to reach you at work."

So, I let him suffer a little longer as I slowly went into the kitchen and sat at the table. I asked him if he could make me a cup of tea, and then I could tell him more about my conversation with the doctor. I tried to look as pitiful as possible.

"Husband," I said, "He confirmed that we are definitely pregnant. However, he said his test showed that we could be having four babies."
As the blood rushed out of his face, he asked,
"What did you say Victoria? Did you say four babies? Four babies?"
I decided to stop teasing him and said,
"No. For real Husband, we are not having four babies, but, we are having twins."
As I sat grappling with the fact that I actually said those words out of my mouth, my husband was ecstatic. Once he calmed down, he had a million questions and still couldn't get over the fact that his actual dream, of having twins, had come true. He wanted twins because they ran in his family. I could tell already, he was going to be a great father.

1983

Once we accepted that we were having twins, we contemplated all of the variables with twins. Would we have identical twins or fraternal; girls or boys, a boy and a girl; or if they were identical, would one have gray eyes and the other have brown eyes (My husband's father had gray eyes). At least then, we would be able to tell them apart. We anticipated having fun with the babies.

My pregnancy went very well. As I grew, the happier I was. However, by the time I got to my seventh month, my neighbor, hated to see me coming.

He would always say, "Lady, you haven't dropped that load yet?"

I would always say, "No. no, I am still baking."

His response would be, "You need to turn the oven off."

I would just laugh. I didn't care how big my stomach got. I wanted the world to know that I was pregnant.

Stopping traffic did not necessitate the use of a police officer or a traffic sign when I was around. All I needed to do was step out onto a curve, and I would cause a mini traffic jam. Traffic would stop in all directions. I would receive cat calls like,

"Are you ready to drop that load?" or "Baby Girl, you need to go somewhere and sit down."

Around that same time, I commented to the doctor that, when I drove, while sitting at a traffic light, with my foot on the brakes, the steering wheel would move back and forth. I teased that the babies were trying to drive.

He said, "That's it. No more driving for you, young lady."

By the eighth month, I was walking with the assistance of a cane, because I had so much pressure on my pelvic area. When the nurses and doctor saw this, they were done with me.

"Why the cane?" the doctor asked.

I replied, "It is either walk with the cane or fall on my face."

I was happy about my growing belly; however, the last two weeks of the pregnancy were brutal. The closer I got to the end of the ninth month, the harder walking, and sleeping, and standing, and sitting became. During the last two weeks of the pregnancy, I crawled to the bathroom, and came down the stairs on my buttocks. I was so big and miserable I prayed… *God get these babies out of me.*

Ultra Sound of Twins in the womb

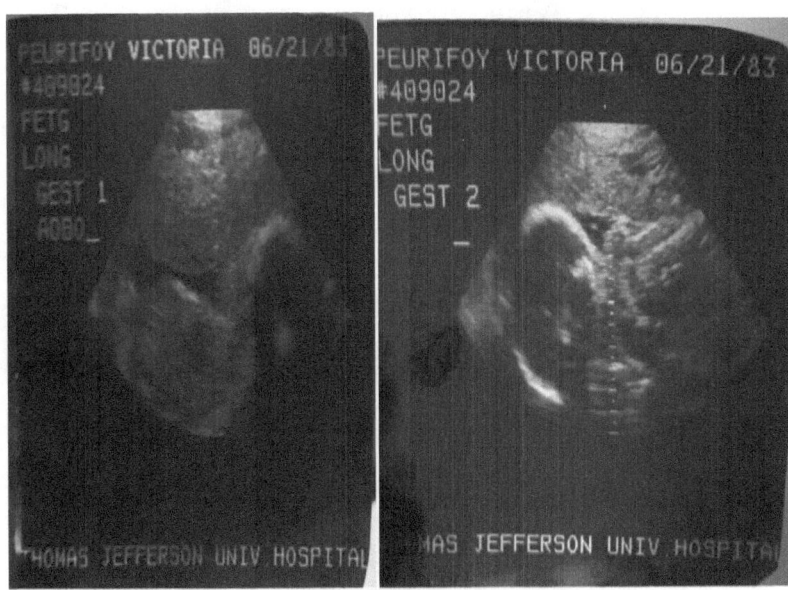

Baby A   Baby B

The Delivery 9-14-1983

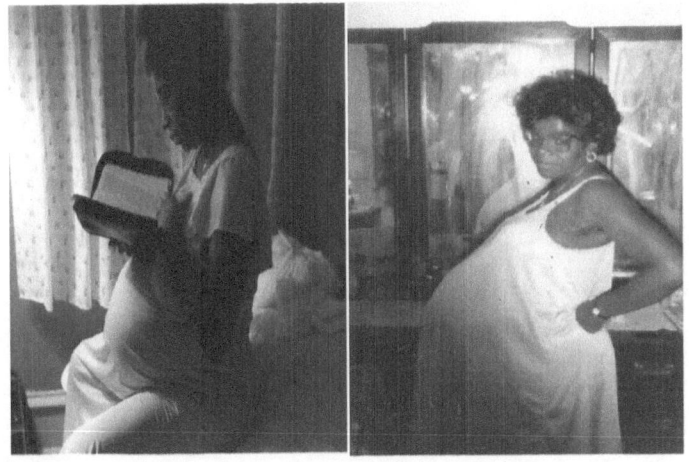

We arrived at the Thomas Jefferson University Hospital ready to deliver the babies. We did not know what we were having. We wanted it to be a surprise. The nurse brought us to the labor room, and proceeded to place the monitor on my belly. Then she gave my husband scrubs to wear, along with a surgical cap and mask. By this time, my husband was really acting like he was *Dr. Peurifoy.*

The nurse showed him how to read the monitor that registered the contractions that I was having. Why did she do that?

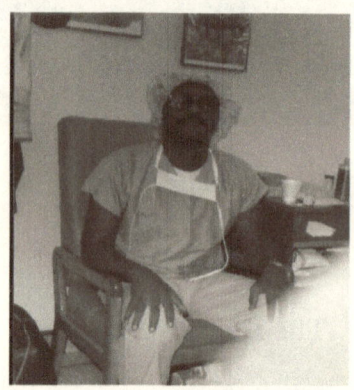

In the labor room, Dr. Peurifoy was saying,

"Victoria, another contraction is coming." Before I could react to his declaration of information, the pain hit me and my eyes rolled up into my head. As the next contraction came, I looked at him and said in staggered emphasis,

This...ought...to...be...you."

My loving husband's response was, "Naaaaw, Victoria, you got this…you are doing just fine."

What God had planned, was for me to carry two beautiful little girls; even if little was a relative expression. Previously, the doctor had suggested that the babies would be about five pounds each. Baby "A" was born naturally at 7lbs 2 and three-quarter ounces. Baby "B" was born by C-section and came in at a whopping 7lbs eleven ounces.

Baby "B" had gotten out of position and traveled back up into the womb; she was exploring her surroundings that she missed while her sister was still on the scene. In the process, she got the umbilical cord wrapped around her neck. She went into fetal heart distress.

The doctor had tried to manipulate Baby B within the womb; to get her to cooperate, but then this doctor, a tiny Asian woman, was literally on top of me saying,

"Victoria, we are going to have to put you on oxygen. Baby "B" is in distress. We will have to perform a C-section."

My husband had cut the umbilical cord for baby "A" and was being ushered out of the delivery room because of what was happening to Baby "B."

Neither I nor the doctors could believe that I had 14 pounds plus of babies. I had only gained 26 lbs. during the pregnancy. However, upon delivery, I lost 40 lbs.

Who do they look like?

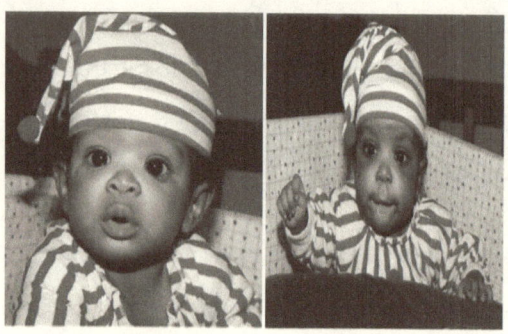

| **Baby A** | **Baby B** |

The twins were born September 14, 1983

My husband was professing that the girls were identical twins. The nurse brought both girls to me and I immediately saw that they were not identical. One was fair, one, brown, one had eyes like her daddy but eyelids like mine, and the other baby had almond shaped eyes just like her daddy. One had a skinny face; the other had a round face. They were confirmed to be fraternal. I jokingly called them my brass monkey babies. However, to me, the great expectation turned out to be the joyous delivery of beautiful girls. For what more could I ask? Thank you, God.

Birth Announcement

THE CREW HAS ARRIVED!

TWO GIRLS

BORN: September 14, 1983

WE'VE CHANGED OUR FAMILY NUMBER,

IT'S INCREASED NOW BY TWO.

WE'RE DOUBLING ALL OUR ORDERS,
OUR JOY IS DOUBLE TOO !

NAME : NICOLE   NAME: CHRISTINA

3:11 p.m. :TIME: 3:30 p.m.

7 lbs 2¾ oz - Weight: 7 lbs 11¾ oz

PROUD PARENTS: ARCHIE & VICKY PEURIFOY

PRAY FOR US AS WE REJOICE !

This page left blank on purpose

III. A Pain in the Neck

(Cervical Laminectomy)

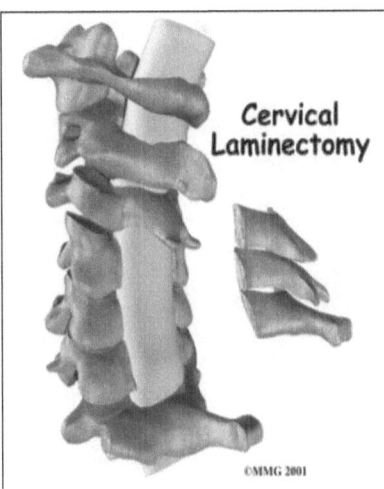

"He who is not courageous enough to take risks will
accomplish nothing in life."

Muhammad Ali

This page is blank on purpose

Dorothy Bernard once said, "*Courage is fear that has said its prayers*" I would need courage for what I would soon be told.

"Do you know that you lean to your left when you walk?" Dr. S. Kind asked.

I said. "No!"

He affirmed, "You do." he affirmed.

"I will need to order an MRI, and an X-Ray of your spine." To myself I think, *now what's happening?*

From late 1988 – spring 1994, I suffered with pains in my shoulders and especially my neck. At times, I felt like my neck was actually hanging so low to the left, that my ear touched my shoulder! That's probably impossible to do; so I'm exaggerating, but that's how it felt. Let me clarify.

In 1988, we purchased a new home that was a fixer-upper. It had beautiful natural wood paneling, flooring, steps, banisters, window frames, seals, baseboards, and crown molding… all of which I loved. However, the walls in the entire house were covered in awful wallpaper from the 1940s. I could not stand it. I read some books on taking down wall paper. It didn't look difficult.

I needed to cover the base boards and floor edges. In addition, a lot of newspapers would be needed on the floor for walking and avoiding slips and falls. Dif® wallpaper remover, a wall board knife, a bucket, a solution sprayer, a sponge, and a perforating tool would be needed to do the job correctly. The instructions were simple. My husband had a bad back. So our conversation went like this,

"Baby, I can do this. It's not much different than painting. Just a few extra steps…a piece of cake."

All of my maladies began in 1989, right after challenging myself that I could take wallpaper off a ceiling in my kitchen. The wallpaper was as old as the house which was built at the turn of the 20th century. The ceilings in my Victorian home were high. As I stood on the six foot ladder, I took the paint roller and dipped it into the solution for removing wallpaper. I stretched the rod and roller much farther than I should have. Rushing the process I guess. I heard something in my neck that sounded like a nut being cracked. It could have been my imagination, but at that point, I stopped and slowly descended the ladder. The ladder was left in the middle of the floor.

I didn't immediately go to the hospital or doctor. I thought if I soaked in Epson® salt and put some heat around my neck, I would be fine in the morning. However, my husband noticed I seemed uncomfortable when I sat too long and when I walked.

He asked, Victoria are you alright? You seem to be leaning when you walk.

I responded, I am a little uncomfortable.

Initially, I brushed it off as something that would go away with time. However, for years I suffered and progressively got worse. The prescribed Tylenol® and neck brace were not correcting the problem; in addition, my thoughts about having corrective surgery were voluminous. My doctor had talked to me about it, but I didn't want to be cut open like a piece of meat. I was afraid of the unknown, but I wouldn't admit it. I even thought if I went to a chiropractor he could crack me back into shape. However, my general practitioner told me that I couldn't see one because he might do more damage than good. My husband had told the children that I would need help sometimes. He wanted them to know their help was important. My husband and the children simply assumed the role of helping me whenever I needed it.

As time went by, more physical problems occurred. I left one doctor, to a change of doctors, to my final new general practitioner, and subsequently a specialist who spoke of surgery.

All I ever thought of was the potential for becoming paralyzed. In my stubbornness, I refused to consider an operation; even though the problems were compounded by the stress of my suffering. When I sat in a high back chair that had arm rests, my falling asleep in such a chair left me incapacitated. Falling asleep locked my chin to my chest and left my arms immobilized. Upon awakening my eyes would look up and around, but my body would not, I could not, move. It was so bad, my 10 year old twin girls and six year old son had learned how to move me when my husband wasn't available.

"Come on Mommy,"

My son Zee was saying, as he took his small hand to my chin and gently raised my head so my neck could straighten and I could see everything in front of me. The twins, Nicole and Christina, were on either side of me, gently moving my arms off the chair's arm rest.

"Mommy, we're going to move your arms now, so they can hang on the sides of the chair."

The next part of the process was to get me to the edge the chair so I could stand. The girls were on either side of me. Literally, they both pulled me to the edge of the chair and then wedged their shoulders under my arm pits to hoist me to an upright position. They would always say,

"Mommy, don't move, just stand still until the blood flows through your body."

They probably remembered me saying,

"I have to allow blood in my body to flow when I stand up."

This ritual went on for over two years.

I always slept well, but time would take its toll on my husband whose ritual each morning was to assist me. To get me out of the bed, he would use his right hand to cup my right hand. This allowed him to swing me to the right and bring me to an upright position. I never could move right away. I always had to let my body regroup from lying down. Afterwards, I could stand and walk.

As the months passed, my condition worsened. My husband had to help me get out of bed because once I fell asleep, I couldn't move. One morning, he moved me the way he always did, but on this particular day, something went wrong. As he went to swing me around into a sitting position, I experienced excruciating pain that shot from my tail bone to my head. I screamed so loud, that it scared him.

"Victoria," he said, "That's it. You have to take care of this because I hate seeing or hearing you suffer. You have got to talk to the doctor again about how you can resolve what's going on in your neck."

At one of my doctor's visits, my husband had heard the doctor say that I had been treating my neck condition conservatively, but I would have to stop denying the inevitable. Something more would be needed. My husband told the doctor that he had talked to me about it, but was never able to convince me to have the operation.

Often, I would say to him, "You're not the one getting cut open like a cantaloupe. I just don't want to be cut again. I've had enough operations and I don't want another one."

For several years, I had worn the neck brace which gave relatively minor relief…until I took it off; then the shooting pains would begin again. One day, while helping my friend Lee with her tax return, without my awareness, she was focused on my neck.

Out of nowhere, she asked, "Where's your neck brace?"

"I took it off…it was getting on my nerves." I said.

Her response, "When do you plan to have corrective surgery?"

I said, "I'm not having any surgery…they're not cutting my neck." Her response was the catalyst that propelled me into more anxiety.

"What color wheel chair do you want? I'll buy it for you. You *will*, and I mean, you will *definitely* be in a wheel chair within four to five years, without the surgery," She proclaimed.

It's funny when I recall how I had known this young lady since she was a baby. Now here she was a nurse who had moved up in rank at her hospital of choice and was telling me what would happen if I did not have surgery. This particular friend was an operating room nurse who had seen it all, so she spoke with knowledge.

Then she said, "Look, if you change your mind, I can give you the name of a good doctor."

As tears streamed down my face, and my husband listened, to the new doctor, Dr. S. Kind, gave my prognosis.

"Victoria, in order to correct the problem, we will need to perform a cervical laminectomy. The other option is to go through the front part of your neck.

I thought to myself, *I really didn't want that. The idea of cutting my neck doesn't set well with me.* So, I agreed to the first option.

He said, "We will open a number of your vertebrae, the area between the discs, to allow the cervical cord room to move away from the disc compression. I had plenty of questions.

I made inquiry, "How many of these surgeries have you performed? How long have these surgeries been in practice? How long does the surgery take? How long will I be in the hospital? What's my recovery time?"

All in one breath, my questions ricocheted off the walls, and he returned rapid fire answers.

"Over 300, this operation was first done in 1887 by a surgical professor from the University of London, named Victor Alexander Haden Horsley. The surgery is about 6 – 8 hours. Once you have the operation, you will be in the hospital several days but be aware, it will take you months to heal. You will also go through physical and occupational therapy. It's going to take time."

Day of the Operation

As I lay on the gurney, on the second floor of Germantown Hospital, in the spring of 1994, my mind was racing.

My husband was comforting me, "Victoria, you will be fine. Dr. S. Kind is a good doctor or Lee would not have recommended him. I like how he explained everything; so don't worry."

Once in the operating room, everyone started moving briskly.

"Mrs. Peurifoy, I am your anesthesiologist, I want you to breathe normally and count backwards starting at 100

"OK," I say, "100 – 9."

The Anesthesiologist knew I would not make it to 99.

The doctor was giving instructions, "Vicky, Vicky, wake up. Wiggle your fingers." As I looked at the doctor, I smiled.

"Vicky, wiggle your toes." I smiled at Dr. S. Kind again, obliging his request. I recall seeing lights flashing overhead as I was whisked off to CAT scan and MRI. I don't even think I had a concern at the moment; I was so drugged.

I found out months later that Dr. S. Kind was scared to death after my surgery. When he was giving me instructions, all I was doing was smiling at him. However, I *did not* wiggle a finger or a toe. The emergency CAT scan and MRI were to ensure that there were no blood clots. There were none. As a result, he gave me a mega dose of *Flexeril Cyclobenzaprine®*, which was a muscle relaxer. It allowed me to do as I was instructed. I was able to raise my right index finger an eighth of an inch off the table. My movements were in slow motion, but he was happy. I was too drugged to care.

Panic

Normally, my husband never responded well to crisis. He was in a panic. His nerves were inflamed and he couldn't think straight. His first thoughts…call Aunt Dar. My sister had always been my backbone. When I married, she became my husband's backbone…next to me.

From our home, my husband called my sister Dar to say,

"Aunt Dar', (that's what he called her) I can't find my wife."

"What do you mean, Brother-in-law?" She asked.

"They keep moving her from one place to another, I can't find her."

Dar responded, "Brother-in-law, I'll call you back."

Dar went into business mode. She called the hospital. As the operator answered,

Dar said, "Hello, my name is Dar Huggins. I would like to be connected to the operating room for Dr. S. Kind, who is operating on Victoria Huggins Peurifoy." The operator said,

"Thank you. Just a minute"

Dar was connected to the operating room and someone picked up the phone.

"Hello, O.R.," a voice answered.

"Hello, I would like to speak to someone with some intelligence concerning Victoria Huggins Peurifoy. My name is Dar Huggins, I am her sister. I am trying to help her husband, because he is in a panic. He cannot find his wife!"

A familiar female voice answered "Hi Dar. This is Lee."

Dar exclaimed, "Thank you Jesus!"

Lee explained that I had been in recovery; but was having a CAT scan and an MRI done after the operation. Although I was awake, I was not responding to instructions or stimuli.

Lee said, "I will call you back when she's in the room and I will call her husband too." Lee assured her.

Although Lee knew that I was having the surgery that day, she was not aware, until arriving to work that morning that she had been assigned to my case as the assisting nurse.

Lee told me later that she was happy she got the assignment. She never let on that she knew me in the operating room. She said, she felt empowered because she could facilitate as an advocate for me and my family. She was never afraid for me because she had faith in the surgeon. To her, it was God's divine intervention that she was there and I agree.

It's Therapy Time

I spent three days in intensive care, five days in a regular room, and 45 days in Evergreen Rehab Facility, in Chestnut Hill. My recovery took one year of in-patient and out-patient care. I had to re-learn how to walk and use all of my limbs; I walked like an unsteady toddler. I had to learn how to move my upper and lower limbs, to write, sew, take pictures, and do simple tasks. When I arrived at Evergreen, a nurse's aide greeted me and my husband at the ambulance. The wheel chair that she brought with her had a high back; that would support my neck and head. After the surgery, my neck was so swollen that I still had to wear a neck brace. In addition, while in that wheelchair, I had to wear a seat belt; so that I didn't move much.

As we entered the facility, I spotted a humungous receptionist desk, and wheel chairs everywhere. There was a sweet aroma in the air…like lavender. All of the patients were either in wheelchairs, or walking with walkers or canes. I couldn't even steady myself on a walker and forget a cane. In the hallways I noticed the walls on either side had rails for the patients to hold.

My first meal would not be in my assigned room, but rather in a group dining hall. I was wheeled to the table by the nurse and left there with other patients. Some of them had broken legs, neck braces like mine, immobile shoulders, and other maladies about which I did not inquire. My special wheelchair was comfortable for me, but feeding myself was a challenge. I couldn't lift a fork to my mouth. A nurse's aide was kind enough to assist me as my salty tears met my food and my tongue.

She kept saying, "You'll be alright, give yourself some time."

The physical and occupational therapists were going to be my life blood. The goal was to teach me everything possible to get me back on my feet again. The physical therapist, Phil, said,

"I hope you are prepared to work hard. We are going to get you back to your old self."

Unfortunately, on the first day I was in therapy, I fell to my knees on the bridge-like walkway that had handle bars on either side. In front of this bridge were pallets where patients exercised. I was trying to focus on the pallets because I saw other patients lying down on them. I really wanted to lie down too so I could go to sleep. However, my legs were too weak to walk with a walker, and my wrist was not strong enough to push my body forward with the walker.

Phil asked, "Are you trying to give me a heart attack?"

"No, but if you are so inclined, knock yourself out." I retorted. He laughed at my retort, and helped me into my wheelchair.

The therapy room looked like a children's wonderland. There were different size balls, ropes, weights, bikes, and colors…lots of colors. The colors all represented various levels of difficulty. A child would have fun with the balls, ropes, bikes and even the puzzles. The occupational therapist found out that I had been a seamstress, so when she felt I was able, she had me sewing. All of the patients at the facility had brocade tote bags attached to wheelchairs and walkers. I would be making these tote bags. She was checking to see my endurance and what I could handle as I sat at the machine. She also had me playing with special balls, colorful straps, puzzles, writing, and using other apparatus designed to strengthen my arms and hands.

A Quiet Storm

Two weeks into therapy, the quiet in my storm came with the assistance of an angel. God confirmed my belief in Him when an angel came to visit me one night. I was lying in my special bed, when pain began to travel up my neck and continued to my brain... then back down to my feet. The ability to push the button seemed virtually impossible. When the nurse appeared, all I could say was,

"Pain!"

She asked, "Did anyone tell you that before you go to sleep, to ask for your pain meds?"

I said, "No!"

She gently helped me move; placing her hands and arm behind my head and back, helping me move to an upright position. She let me sit on the edge of the bed for minute, administered my medication, and then helped me to the restroom. Her name tag said Angeline. I thanked her for her help.

The next morning when I asked about Nurse Angeline, the nurses said they didn't have a nurse on staff for day or night shift named Angeline. I was awestruck! God sent me an angel in disguise. I remembered a glow all around her; but I thought it was my imagination. I wondered how she moved me without hurting me. I know now that God was with me that night and every night and day of my recuperation. I love the angel he sent.

Nurse Angeline

Spiritual blessing always abound.
Never do we expect what comes our way.
Glows of mercy touch our hearts each day.

Nurse Angeline, who no one knew,
treated me with care. She exude such grace
and mercy too. An Ordinary appearance at first,
later she was deemed extraordinary.

She was unknown to anyone, but her presence
blessed my heart with her gentle hands.
Nurse Angeline, may God forever bless you…for
blessing me.

© February 1, 2017

I knew from that point on, that regardless of how difficult my recovery would be, I would weather my storm…until…

Depression

It had only been three weeks, but my mind was rushing the rehabilitation process. Fear was creating irrational thoughts such

as *I'm never getting out of this wheelchair; rehab will take forever.*
Depression was trying to get me. As I sat in my room, I began
worrying about my job. I had told them I would be out for two
weeks. The weeks were piling up. I was being paid my regular
salary through sick leave, donated leave, and hours from our
leave bank. Yet, fear was ravaging my spirit. I was questioning
whether or not I would ever be able to go back to work.

As a tax auditor for the Federal Government, I needed to be
able to write and document cases. That would require the use of
my hands, neck, eyes and conscious mind. I was conscious, but
immobile. Finances were also on my mind. My husband and I
always pooled our resources, but I made substantially more than
he did due to the recession and the crash of the steel industry. If
I couldn't go back to work, what would we do?

As my nurse's aide walked into the room, with her usual jovial
high pitched voice, she said,

    "Good morning Mrs. Peurifoy. How are you today?"

I didn't say a word. I actually grunted at her. She looked at me
and walked out of the room. Two minutes later, she came back
into the room, closed the door behind her, pulled up a chair next
to me and said,

    "I don't know what's wrong with you this morning, but
    you better get it together. Do you know that the nurses

record everything you do…or don't do in this room? If they come in here, and see you like this, and you grunt at them the way you just did me, do you know what they will do?"

"No!" I exclaimed.

She continued, "They will call the Psych doctors, and then the doc's will want to give you drugs!" She repeated, "You better get it together."

She pushed her chair away and walked out of the room. I thought long and hard and quickly about what she said and gave up my pity party. I knew for a fact that I didn't want any more drugs if they were not absolutely necessary.

Recall

Recently, I asked my youngest twin daughter if she remembered me going through all of this. She said,

"Mommy, I just recall that Daddy and Uncle Harold sat us down, and told us that you were very sick. Our home life and flow of the house was going to be different. Daddy did his best to fill the gap. I remember after your surgery, neither of us knew what we were supposed to say, feel, think or otherwise. When Daddy took us to visit you at Rehab, we all (the children) talked about how much slower you seemed. While there, the therapist taught us how to help you! Mom, we were in our zone at the time…We were just being kids. You being sick and almost helpless did not help. Subconsciously, we all were worried. When you came home, everyone that could or would just pitched in and helped. That all I recall."

Forging Ahead

A long time co-worker and friend came to visit me. As we sat talking, my subconscious mind was also talking to me. It was saying, *you can do it, you can do it*. I am sure, that my friend thought I was losing my mind because I picked that moment to try something that I had been working on all week. I was desperately trying to make my trembling index finger point toward my nose. With nervous and shaky anticipation, I continued moving my right index finger upward toward the mark. Meantime, my friend was still talking to me yet watching me, not understanding what I was trying to do…and I didn't tell her. All of a sudden, my eyes crossed, I hollered.

"I did it! I did it! I touched my nose…I touched my nose. You don't understand. I've been trying to do that for a week."

My friend smiled at me, and tried to understand. She probably was thinking, *my girlfriend is going crazy*.

My church family sent me greeting cards wishing me well. But in addition, cards were coming from choir members, church auxiliaries, co-workers and anyone who knew that I was in therapy. They also visited me.

During one of those many visit, a choir member who came to see me at the rehab found me in the activity room. I wasn't walking yet, was still confined to my high chair, listening to a patient talking. She observed that nothing was wrong with my voice. I began singing, *Amazing Grace* as I smiled, looking like a mummy sitting in a chair. She told the choir members, she was shocked that I still had a voice, but was encouraged by my strength.

Friends from a second choir I sang with visited me as well. They took me for a walk, encouraging me to do all I could to get back on my feet. My husband's best friends even visited. They tried to sneak me a bottle of wine. However, since I wasn't a drinker, I told them to save it for a special occasion. They just wanted to lift my spirits.

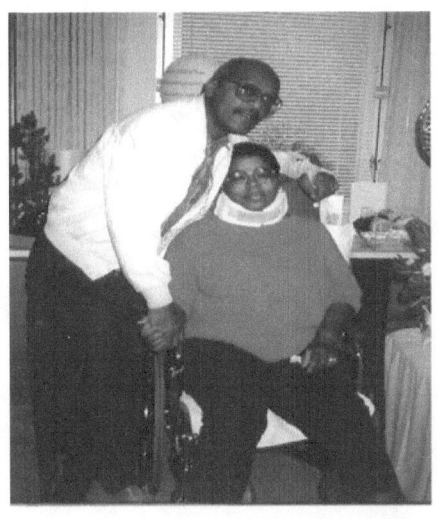

During my therapy, it was determined that I needed bifocal glasses because initially, I could not bend my neck forward. The bifocals allowed me to look down and see the steps in the therapy room, as well as outside. They would come in handy for my next therapeutic adventure. I couldn't drive my car for eight months. I hated losing my independence. While in therapy, I learned why donating cars is important.

One day, the therapist, Phil, said, "I noticed, in your records, that you drive."

Quickly responding I said, "Yes, but I can't right now. I cannot turn my head from side to side."

He said, "We will use heated pads and exercises to loosen your neck muscles; but in the meantime, I have a surprise for you today."

Excited and wide eyed, I asked, "What is it?"

Phil explained, "You are going to practice driving. I need to see how you get in the car and how you would transition from the brakes to accelerator."

Based on what Phil had just said, I walked with the new four prong cane, which I had been given the week before. I headed toward the front door of the complex.

Phil asked, "Where are you going? We're not going outside; we're going to the therapy room!"

When we arrived at the therapy room, there before me, was a car... or I should say a piece of a car. It only had front seats; a wind shield, and windows on the front doors, but no trunk, and no wheels. It was once a real car with its guts removed. I looked at the therapist in shock.

More Shocks

Friends, family and neighbors had been good to me while I was away from home. All I needed to do was call. They would really come in handy with a huge development at my home.

"Hello Mommy, this is Nicole."

She didn't need to tell me, I knew my child's voice.

She continued, "I have something important to tell you. I'm a big girl now!"

I said, "I know you're a big girl!"

"No Mommy, what I mean is I got my period today!"

"What!" I hollered.

"Are you alright? Did you tell your daddy? Do you have your emergency kit I gave you? Aaww, my baby is a big girl now."

She only answered one question. "No Mommy, I'm not going to tell daddy…this is private." She exclaims!

I responded, "Okay. I will call Ms. B; she'll know what to do for you."

I felt so bad that I wasn't at home during this milestone in her life. At 10 year-old, I felt she was quite mature about it all.

I was so busy trying to line things up for my daughter from my hospital room that I forgot to tell my husband. However, he found out.

"Victoria, one of those girls got their menstrual. I don't know which one it is, but I need to talk to her." He said.

I could hear the panic in his voice, and I knew he was trying to be brave. In addition, I also knew there would never be *that* conversation. I shared with him that Nicole had told me it was her and that I was sorry… I forgot to tell him. He was also informed that the neighbor was going to take her to the store. To say the least, he was relieved.

Going Home

As I was coming to the end of my therapy at the rehab, I was informed that before I could be released to go home, they would need to go to my house and to my job. The purpose was to determine what apparatus I would need. They drove me in a van especially equipped for handicapped patients. I didn't consider myself handicapped, just a little challenged. When we got to my home, they counted my steps inside and outside of the house. They looked at my bathroom, my bed in my bedroom, and they observed how my kitchen was situated. At my job, they examined my cubicle, the desk and even the distance to the bathroom. The bottom line was that when I returned home, I had grabbers to help with reaching for things in the cabinets, and devices to help me button my shirts, put on my socks, and shoes. They even gave me an especially long shoe horn. All of these gadgets were a tremendous help. Most of these items were made with plastic and sometimes they had metal on them. For my job, because I was always writing, they gave me a special table designed to sit on my lap or on my desk. It reminded me of a breakfast-in-bed table but it didn't have legs or sides. It had levels, so that I could adjust it to the level that was comfortable for me. They even gave me a special screen to cover my monitor so I would not strain my eyes. When I returned to work, for four months, I was only permitted to work part time so my body could acclimate itself to work life again.

By the time I was released from outpatient physical therapy (almost 11 1/2 months later), I could only drive less than a mile radius. It was six months after I was released from therapy before I was able to drive the Schuylkill expressway. The first time on it was scary, but I was on a mission. My husband was letting me drive to church. I was proving to myself that I could do it...and I did.

This page left blank on purpose

My Cervical Laminectomy Scar today

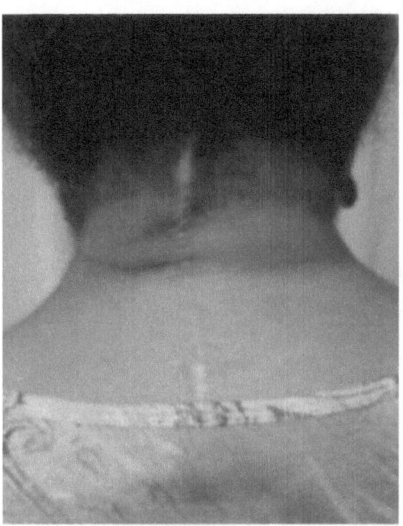

In-patient and out-patient care

My inpatient and outpatient experience at the rehab was wonderful. As much trouble as I gave them, moaning and groaning about the exercises they wanted me to do and wanting to go home, knowing I was not physically or emotionally ready, the entire staff of doctors, nurses, and therapist, were great. They did their job. When they finally released me from outpatient services, Phil said,

"I forgot to tell you that you don't need that cane anymore. All you are doing is dragging it behind you anyway.

However, still take it home, because there may be times, when you *will* need to use it again.

I thanked them all for their kindness toward me, even Phil, my physical therapist who tortured me the whole time.   Even he smiled when I gave him flowers on my release date. When I got home, I called my 89 year aunt and told her I had a cane for her. She gladly accepted it.

My life after surgery, after therapy, after going back to work, for the most part was normal. The problems I have now are weather related. Cumulus clouds and over cast days bear down on my shoulders and neck. When this happens. I take a pain killer and rest. Barometric pressure is real. That's what you feel…intense pressure. When a dentist is bearing down on your jaw to remove a wisdom tooth, that's how it feels. My grandmother used to say,

"Chil' when the weather is getting ready to be bad, you can feel it throughout your whole body before the weather even gets here!" Now I understand what she meant.

IV. I Can't See

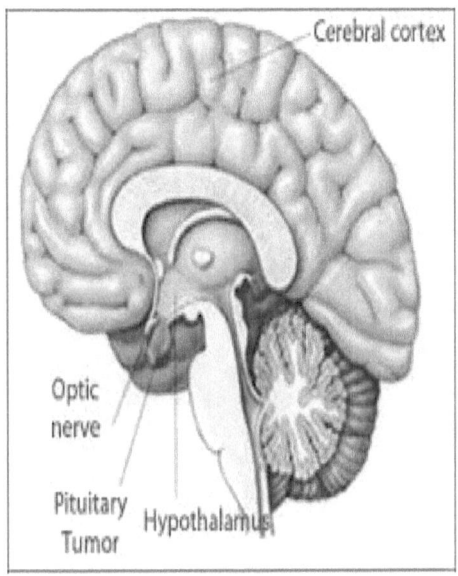

Source:

http://www.northernendocrine.com.au/pituitary_20.html

"When a woman is talking to you, listen to what she says
with her eyes."

Victor Hugo

This page is blank on purpose

Why won't you believe, *I CAN'T SEE!*

Working on prom dresses can be very tedious. I actually loved what I was doing. Sewing was one of my favorite hobbies. However, I was having trouble seeing how to thread the needle of my sewing machine. My daughters, who were in the 12th grade at the time, would have to come and thread the needle for me. I could not see the opening of the needle. My daughter, Nicole, had purchased fabric that was called *pony hair*. It was soft and smooth to the touch. In addition, she had to use soft netting for the under panel of the dress. I told her and her sister that they would have to help me with the dress; because, I couldn't see what I was doing. At first, I thought my daughter just wanted me to place the pony-hair onto the netting as a whole panel.

However, she said, "No, Mom, I want the designs on the fabric cut out and strategically placed onto the netting." There was no way I could do that without help.

It was shortly after this episode of trying to sew, when I began to notice that I seemed to have lost some of my sight. My vision had become so bad that if I sat at the computer, I had to take my glasses off; then put my glasses back on. Sometimes I almost put my nose on the monitor because everything looked blurred. When I was driving, my peripheral vision had gotten so bad that I actually had to turn my body around to the left or to the right to see if there was anyone on the side of me because the side view mirrors did not help.

Trip to the Optometrist

The doctor was asking, "What is the problem? You've been here two or three times in the past 6 months and your vision has not changed; at least not according to the test that we've done."

In frustration I said that, "I really don't care about the tests you've taken; that have kept saying that my vision has not changed. There must be another test you can give me. There has to be! I'm telling you I can't see. You're telling me about some tests which are not effective enough to determine the problem."

He finally said, "I'm going to set you up for a Field of Vision test (F.O.V.)."

I can't see

What color is the sky?

Is it gray or is it white?

Birds fly right by

I don't notice them in flight.

Children play, but I only hear

their shouts of glee

Sewing used to be fun,

but now it's a burden.

I can't see the patterns nor

the colors of the velvet.

I have expressed my frustration,

which has gone on deaf ears.

I can't see, write, or read.

but no one believes that I can't see

© *February 22, 2017*

Field of Vision

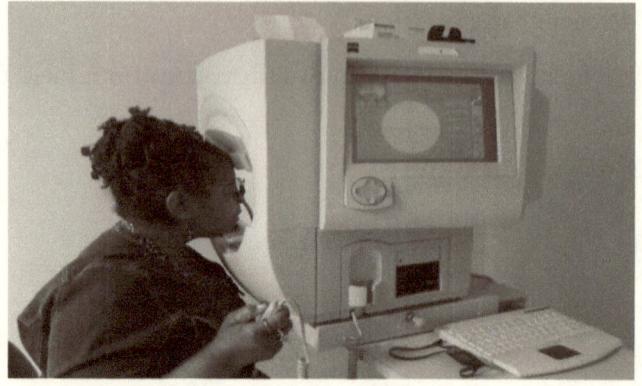

Of course, I wanted to know.

So I asked. "What is a F.O.V. test? What does it do?"
How does it determine the problem with my sight?"

The doctor said, "The Field of Vision test is designed to
determine what you can see as far as light flashing at you.
It's not a difficult test. Your head will be facing a dome
that has what looks like little twinkling star lights in it.
Initially, the inside of the dome will look as if it has many
tiny holes or like little dots throughout the sphere.
However, there will be light that shines through the dots.
Every time you see a light… you push the button on the
hand wand which you will be given."

So, I agreed to the testing because maybe, I hoped, finally, I
would find out what was wrong with my sight.

I got plenty of rest the night before I took the test. I wanted to be alert and aware of everything that was going on around me as I was taking the test. A young lady greeted me saying she would be giving the F.O.V. Then she told me that I would be sitting in the chair facing the dome.

She explained, "Every time you see a light flicker you should click the button"

I said, "Okay. That sounds easy enough."
The test took about 40 minutes. I felt a bit tired after the test because it was somewhat monotonous and it tended to make me drowsy, but I forced myself to stay alert.

Test Results

While I waited, the doctor was reading the test results; the nurse came to me and said, "The doctor will see you now."
As I entered the doctor's office, he invited me to take a seat and he began to ask questions and saying things that made me respond with knee jerk reaction.

With impassioned concern, he said, "Victoria, according to this test, you've lost 60% of your vision...*you can't see!*"

Agitated, I retorted, "I've been telling you for six months that there was a problem. I don't know why you wouldn't believe me when I kept telling you that I couldn't see!"

He responded as if his concern took precedence. He went on about how I could be declared legally blind and that he would have to notify the state that my driver's license should be revoked because of my sight.

Immediately, I went on the defense, "Wait a minute, hold up, I drive for my job…my eyes are my livelihood… you're trying to take my ability to work away from me. I've been complaining for months and now you telling me that I'm legally blind. Can anything be done about this?"

The doctor began to lay out a potential plan.

He said, "You'll have to see a neurologist. He will have a different set of test that he'll run to try to see why this is happening. Something could be happening in your brain that's affecting your sight. You will need to make an appointment right away. Otherwise, I will have to contact the Pennsylvania Department of Transportation."

All of a sudden, there was this frantic need for me to get something done and I had no idea what that something was going to be.

Tell Hubby

When I went home that evening, I must have exuded stress; because my husband was asking me what was wrong. During the day, I had time to let my imagination work overtime. *Suppose I have cancer. What will I do then? How much time do I have to live? What about my children? What about my husband? How will he handle the news? I was driving myself crazy...internally.*

"Hi husband. I had that test that I told you about. Now the doctor believes me...after all this time. But I have to see a brain surgeon, specifically a neurologist. They're going to be playing around in my head and make me crazier than I already am."

My husband looked upset to me, but he tried to stay calm. "Victoria, stop playing around. This is serious. When do you have to see the neurologist/brain surgeon?" he asked.

"I don't know, I have to make an appointment...will you go with me?"

He said, "I will go with you, but in the meantime, I don't want you to start with that vivid imagination of yours. You know how you are."

It was too late for that advice!

The Neurologist

The neurologist was able to see me in a matter of days. With my husband by my side, the doctor explained possible reasons why my vision was leaving me. He also talked about the fact that there is corrective surgery for what he believed was wrong. However, if I decided *not* to have the surgery, I would eventually go totally blind. How long it would take was unknown. Just the sound of his words sent a chill up my spine and made my husband squeeze my hand harder. He could feel me tensing. I just kept thinking about all the things that I wouldn't be able to do. *I wouldn't be able to sew. I wouldn't be able to photograph life. I wouldn't be able to write. I wouldn't be able to see my face. I wouldn't be able to tell when something was wrong with what I was wearing. I wouldn't be able to see my children, my grandchildren, my siblings, or my world.* I went right into a song of woe that was out of tune. Thank God no one could hear what I was thinking nor hear the song I was singing. The doctor ordered an MRI and a CAT scan.

The Test Results

When the results came back, it was diagnosed that I had a p*ituitary tumor* on my brain that was lying on my optic chiasm (the part of the brain that affords vision.) The doctor said the pituitary gland lies between vital nerves that contributed to the eyes' movement. He explained that the tumor had probably been growing for a while. Had I not insisted on more tests, it was possible that in two to three years, my vision would be completely gone…the damage would not be reversible. Then I got angry all over again about the fact that my optometrist was ignoring my concerns. My husband looked as worried as I felt.

As a result of the tumor sitting on my optic nerve, the specialist decided that I would need to have my pituitary gland removed. At the time, I didn't realize the importance of the pituitary gland nor how its removal would affect me later. However, at that moment, I expressed my concern about having my hair cut off in order for them to do the operation. I was really going through a vanity crisis; and I knew I was.

But, the doctor said, "The way that we will perform the surgery, your hair won't come into play at all. However, I will need you to see an E.N.T. which is an ear, nose, and throat specialist."

"What does an E.N.T. Specialist have to do with my brain tumor?"

Dr. Kay, (as I started calling him) explained, "During the operation, the E.N.T. Specialist will help me to guide a scope into your sinus passage to the site where the tumor is lying."

Shocked, I asked, "Why would he be in my sinus passage?"

Dr. Kay continued to explain, "You see, the tumor is sitting in the frontal lobe of your brain. We will use a **microscope and endoscope,** (we refer to it as a scope) to help suction out the tumor. That is why you won't lose your hair."

As the doctor answered my next question, I began to feel relieved.

I asked. "Doctor Kay, after the operation, will I be able to see…will my sight return?

"Yes, you should be able to see with your corrective lenses." he said.

The Day Before

The day before the operation, the nurse practitioner, Patricia, talked to me in depth about how everything was going to proceed with the surgery and wanted to know if I had any questions. I couldn't think of any questions to ask her. The doctor had already done a fine job of answering my queries.

The Day After

It was about 7 o'clock in the morning the day after the surgery. I was really hungry because I had not eaten anything the night before; so that's where my focus was at the moment. I sat up in the hospital bed, looked around and realized... *I can see again. I CAN SEE!* As this revelation became apparent, I also realized something else... my nose was running profusely. I pushed the call button to get the nurse.

"Good morning Mrs. Peurifoy. How are you today?" She asked. I said,

"I don't really know yet. I can see alright now, but I have this watery fluid coming from my nose… it just keeps running. The nurse looked alarmed but tried to appear calm and said,

"Let me get you something for that. I'm going to call the doctor also."

Dr. Kay came to see what was going on with me. After his examination, he explained that I was leaking spinal fluid. Not a good thing. He called it a *cerebral spinal fluid leak* which can happen; when there is fluid around your brain and it leaks through a hole in the skull bone. He was going to have to put the drainage tube back in my spine…like a spinal tap.

What I wasn't told

I was wide awake when they put that tube in my back. I screamed so loud, I probably scared the entire floor. It was so excruciating; there was no way to even describe the pain. Another thing that happened was that I stayed thirsty all the time and I drank pitchers and pitchers of water. It was as if I were craving it. They had to keep a catheter on me to catch the fluid from my bladder. No one could understand why I was so thirsty. They figured that I would not be able to make it to the bathroom, so they suggested that the catheter be kept in place

.

I didn't know it yet, but I was about to have a new look. For days, I had to wear those thick layers of gauze across my face to catch the fluid from my nose…until the draining had subsided. It annoyed the nurses that the doctor wanted them to save the soiled gauze. They were saying that even though Dr. Kay was annoying, with the things that he required that they do, he was an excellent doctor and I should be happy that I have a doctor who was as anal as he was. Once the draining stopped, Dr. Kay was satisfied that everything was under control. He removed the thick gauze. However, now I had questions about another issue.

Dr. Kay's nurse practitioner came in to see me. By this time, I had begun checking myself, making sure everything else about me was all right; when I discovered a Band-Aid® on my stomach.

I asked her, "What's up with the Band-Aid® on my stomach? I thought you guys were playing around in my head." The Nurse Practitioner looked at the Band-Aid® in dismay.

She said, "I don't know why there's a Band-Aid® on your stomach. However, the doctor will be on his rounds soon to see you, so make sure you ask him about it.

Rounds

Dr. Kay was very straight laced, always very serious, and he hardly ever smiled. He had about 10 Interns that worked with him and they followed him on rounds to see all of his patients. When he walked in the room, all 10 of his student interns walked in behind him and stood at attention like soldiers as he talked to me. When he finished talking I said,

"Hey Doc, what's with this Band-Aid® on my stomach?"

The doctor looked at my stomach and said, "When we took the tumor out of your brain, it left a hole in the spot where the tumor was. We had to use some of your own body fat to put in the hole; so it would blend with the blood vessels in your brain."

Looking at him with my head cocked to one side, I said,

"You mean to tell me, you went into my stomach and took out a little bit of fat to fill a hole? How come you didn't just go ahead and perform liposuction and take the rest of the fat out of there? What's up with that?"

The look on his face was priceless, but then all of a sudden, he burst into laugher. Once he started laughing, like a chain reaction, one by one, his interns began laughing. It was as if his interns had permission to laugh too. I looked at all of them and said,

"All of you are crazy. You don't know how serious I am."

The doctor became serious again, after regaining his composure and said,

"Victoria, we didn't have the authority to give you liposuction. That's not what this surgery was about. This surgery was about giving you back your eyesight. How are you seeing these days?" He asked,

"Dr. Kay, I'm seeing perfectly fine with my glasses and I thank you for that."

The doctor began to explain that in several days I would be able to go home, but that he wanted me to come back and see him in a week so he could check and make sure everything was okay; and administer another Cat-scan and MRI.

He also said, "You will need to be tested every six months for the next five years, just to make sure the tumor doesn't grow back."

Homeward bound

When I got home, although I could see, I think the anesthesia was still in my system. I really felt tired and lethargic; and another issue was I had no appetite. Previously, the E.N.T. specialist had told me that after surgery, I might see a little disfigurement or change in my smile, however, it would be temporary. The problem that I was having (and then again it might not have been really a big problem), but all I wanted to do was drink water, eat watermelon and salads. I really had no desire for any kind of meat or fish or anything else. As a matter of fact, my next door neighbor had made me a plate of home fries with onions, fresh green beans and baked fish. I should have been able to eat what she prepared for me in one sitting, but it took me three different sittings to eat the meal. It worried my husband and my neighbor that I wasn't eating enough, and I was losing weight rapidly.

Several days after I came home, I was sitting on my bed watching the morning newscast. The picture on the screen looked like something out of a movie. My eyes were playing tricks on me again...or so I thought.

It was September 11, 2001 at about 9:00 a.m. What I was watching on the television was a repeat telecast of a jet plane flying into the World Trade Building in New York City. I kept saying to myself. *This can't be real. It has to be a movie.* That is when I began screaming for my husband...he had to see what I was seeing to assure me that I wasn't going crazy. It was as if I had been blind, and this was the first thing I was seeing with my new eyes. Unfortunately, it was all too real. My eyes were working perfectly fine.

Whose Face is this?

I noticed something else; my face didn't look right. I knew there would be some slight changes around my smile, but I wondered what was happening to my face. *Why did it look round like a pie pan?* I called my general practitioner and I told her I needed to see her and she had me to come in the next day. When she saw me, right away she noticed the same thing that I did.

She said, "Your face is really round and your skin is extremely smooth; it also has a little shine to it."
I also told her about the fact that I wasn't eating that much. She weighed me and found that I had lost close to 20 pounds in two

weeks since the surgery. She suggested that I see an Endocrinologist.

The Endocrinologist Visit

I had my sister to go with me to talk to the doctor in case I forgot to ask or tell him something. My sister nick named the endocrinologist Dr. Curly-top. He was a Jewish doctor who had a head full of curly hair. I explained why I was there, the fact that I had just had a pituitary tumor removed, the fact that I had no appetite, and the fact that my face was round like a pie pan.

That's when he said, "Just from looking at you, I believe you have hypothyroidism; but, I'm going to run test to see what's going on with you."

The Results

When the results came in, his suspicions were confirmed. My body was no longer making its own steroid or cortical; which caused me to have the hypothyroidism. This meant that for the rest of my life, I would have to take a therapeutic amount of medications called Prednisone which is a steroid and Synthroid which was to replace the cortisol that my body no longer made. I have been taking these medicines since 2001 when the World Trade Center Buildings came down in New York. In all of that time, only one of the dosages has ever been lowered.

Fifteen years later, my normal smile exists, no more pie face, and I've gained back the weight I lost. With the miracle of the Internet and the ability to research, even though I didn't want to lose my sight completely, I don't know if I would have taken a chance on the surgery. However, I have to say, I'm glad I did.

Another thing that happened was my hair became very brittle. So brittle in fact, I had my son to cut off all of my hair. He kept asking if I was sure. My hair length was below my shoulder at the time. I told him to have no fear. It was perfectly fine for him to cut it all off. After all of my hair was on the floor he said,

"Mom, you look like me now." I said, "No son. *You* look like me."

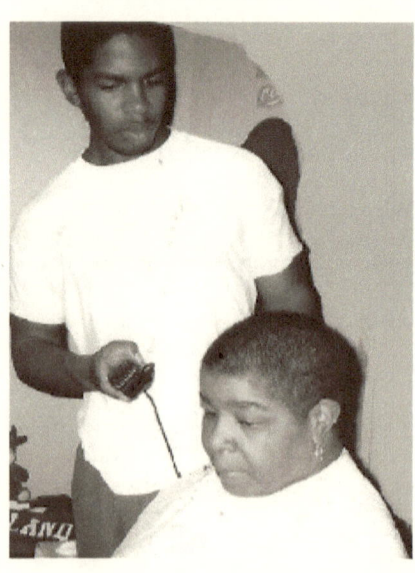

By the way, I completed my daughter's prom dress before I had the surgery. She was the bell of the prom.

Nicole's Senior Prom dress

This page left blank on purpose

V. One Stepper Club

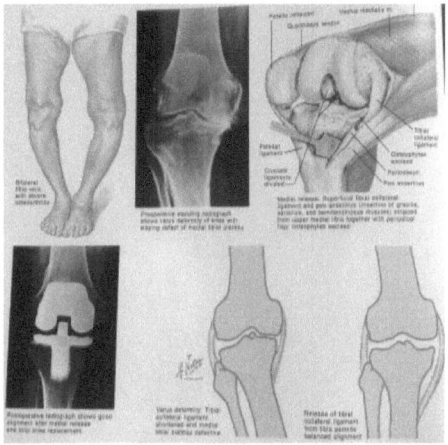

Source:

https://www.facebook.com/TotalKneeReplacement/

"If you can't fly, then run

If you can't run, then walk

If you can't walk, then crawl

But, whatever you do

You have to keep moving forward"

Martin Luther King

This page is blank on purpose

One Stepper Club (2008 & 2010)

"Mom…When did you become bow-legged?"

"Mom, why are you standing in the middle of the driveway?"

These were questions I was often asked by my daughters.

Ten years before, while on my way home from work, as I traveled on Septa's XH Commuter Bus, going to Germantown, I observed a young lady acting strangely. First, she walked from the front of the bus to the back. With each trip up and down the aisle, she would look out of the window, on both sides of the bus. At one point, she acted as if she spotted something. At Manheim and Wissohickon she went to the middle of the bus and signaled the bus driver that she wanted to get off. I watched her hurriedly rush off the bus; dash across the gas station lot, then run into Dunkin Donut®.

In the meantime, on the outside of the bus, through the window, I noticed a young man driving very close to the bus…looking up at the vehicle…as if searching for someone. I figured out that he was who the young lady was trying to avoid or from whom she was running.

The bus turned right onto Manheim and proceeded to the next stop when crushed metal, broken glass and screaming people permeated the silence. Our bus was hit from behind. The same young man, who had been riding alongside the bus, had begun following it and had hit us.

The impact forced me forward as my knees jammed into the back of the seat in front of me; that was covered in a metal plate. Immediately, I felt pain surging through my knees and thighs. Those people who were injured were given a white card to give to a lawyer, if we planned to sue. I was taken to the hospital emergency because I couldn't walk. The doctor I saw, told me my knees were badly bruised, but recommended I stay off my legs, ice my knees, and go see an orthopedic specialist because I would need therapy.

In my 40's, my knees began giving me continuous problems. They would often, ache, swell, and just pain me. I started walking up and down steps sideways; gently putting one foot down onto the step and repeating the same action with the other foot. I had joined the one stepper club! I had gone to the hospital to have my knees x-rayed; losing weight had not worked to ease my discomfort; and calcium pills had not worked. Now that I was in my 50's it seemed the condition of my knees was worsening.

Intermediate Attention

Tylenol became my consummate friend. I had even taken glucosamine chondroitin, which was supposed to be the new miracle drug for folks with knee problems. It didn't work! I had a new doctor to examine my knees.

He said, "Your knees have fluid on them. I will use a large needle to suction out the fluid."

He removed 30 cc's, which is about 1.0143 oz of fluid, from each knee. Seeing the fluid amazed me. The doctor also said that my knees were aging and my cartilage was wearing away. He began to explain a new product that he believed would benefit me.

He continued, "We have a new injection called Synvics®. It can alleviate your pain for up to six months. In some patients it last longer. Would you like to try it?" He asked. Of course, I needed more information.

I inquired, "What is this Synvics® made of? Is it a steroid? How many shots would I have to receive?"

He explained that it was made of a gel-like substance and would act like filler; where the cartilage had worn away.

In addition he said, "You will be limited to three shots over a three week period"

I asked, "Will the shots hurt?"

He said, "You will feel some pressure, but I will numb the area before I give you the treatment. When the initial injection is started, you will have to take it easy for several days."

When he explained that the injection would be much like when he extracted the fluid, I agreed to try the new injection because it sounded like it could work.

Exactly one year after the first set of shots… my pain returned. I quickly returned to the doctor's office. Over the next two years, I was able to get the shots and had the same results.

One day, my daughters and I drove to Staples®, in Chestnut Hill for stationary supplies. We parked in the upper level parking lot, on the right side of the store because the other lot was full. When we walked out of the store, we walked across to the lower level of the lot; the girls walked ahead of me to the car. I had stopped at the lower edge of the slope; because my knees were forcing me to stand in one spot! Both knees were jammed into place. My oldest daughter turned around and saw what I was doing and asked,

"Why are you standing in the middle of the driveway?"

I said, "I cannot go any further. My knees are forcing me to stand here like this. They are locked. One of you has to come back here and get me!"

Last Attempt

When I tried for a fourth time to get the shots, I was disappointed.

The doctor said, "Victoria, I cannot give you any more shots, your cartilage is totally gone. I don't even know how you're walking now."

By this time, I had resorted to using two canes, just to anchor myself. My pain had me walking like I was an 80 year old woman. I was only 56. I had heard that Rothman Institute was one of the best places for hip and knee surgery, so I made an appointment with Dr. K., who was one step under Rothman himself. I liked his demeanor. He was clear and concise. He went over my x-rays and explained them in detail.

"Normally," he said, "I don't like to do knees of individuals as young as you; but you won't be able to walk independently without a cane or possibly a walker. You need the surgery now. The reason why you have become bow- legged is because your knees are sitting on an axis."

He dispelled one of my misconceptions. "Do you really have male and female prosthetics for knee replacement?" I asked.

He laughed and said, "No. There are not. However, there are various sizes of prosthetics; because we learned that one size does not fit all. Think of the fact that we all wear different shoe sizes, that's how we have to look at the size of prosthetics that we use."

I wanted to have both knees done at the same time, but Dr. K. would not hear of it. He felt that the recovery time would be too long. He suggested that I have one knee done at a time, so we opted to do the right knee first. After a long discussion, he had me schedule the surgery. I asked him about recovery and pain, and he had me look at a video about how knee replacements are done. In the meantime, he showed me the prosthetics and discussed how post-surgery and therapy worked.

25 Years

25 years have passed; my son's memory faded,
my cane he remembered giving me; for knees that
knew nothing but to
Hurt me...
Pain me...
Suffer me...

Stairs were not my friend, but he would give me his
six-year-old hand saying, "Mommy, I'll help you
up the stairs, down the stairs, or
around the corner."

Carpal tunnel made pill boxes unfriendly,
but his nimble fingers easily filled colorful compartments
that his Mommy couldn't.
"You took a lot of pills back then."
He recalled.

Doctors were as close as a brother or sister
saw them so often.
Today, he will not take prescription drugs.
But, surprisingly,
he remembered...
the pain,
the cane
the pills, and
the doctors,
25 years later.

@February 11, 2017

Surgery

I don't really recall going under the anesthesia or being in recovery. I know that I woke up in my room with my leg in this contraption. It looked like a boat that had lost its bow. It was covered in lamb's wool for padding. I noticed that this device kept slowly moving back and forth; but I hardly felt it moving.

The doctor walked in the room and said, "I'm glad you're awake. How do you like our Continual Passive Motion Machine... we called it the CPM?"

My response was, "I don't know how I like it... what's it supposed to be doing?" I asked.

He said. "It's designed to keep your leg elevated; pump the swelling out of your knee; helps reduce the pain, and it stimulates your circulation."

Surprised I said, "Wow, is it going to help me to walk too?" I asked.

Later in the afternoon, the same day of the surgery, after I had seen the doctor, the physical therapist walked into the room smiling. She was carrying a silver walker.

I looked at her and asked, "Hello. What's up?"

She said, "Are you ready to walk?"

Quickly I said, "No!" "I don't think this railroad track that is up and down my leg, wants to bend or walk."

She proclaimed, "Well, let's just see about that!"

She walked toward my leg, removed it from the CPM, as I hollered from the movement of the 25 staples that were above and below my knee.

"Ooh, I'm sorry." she said. "I was trying not to hurt you. Before we continue, I will have the nurse to give you some pain medicine. I will be back in 15 minutes."

In that time frame, I had enough time to recover. My leg was so heavy from the swelling that it looked and felt like a fat walrus asleep on a rock.

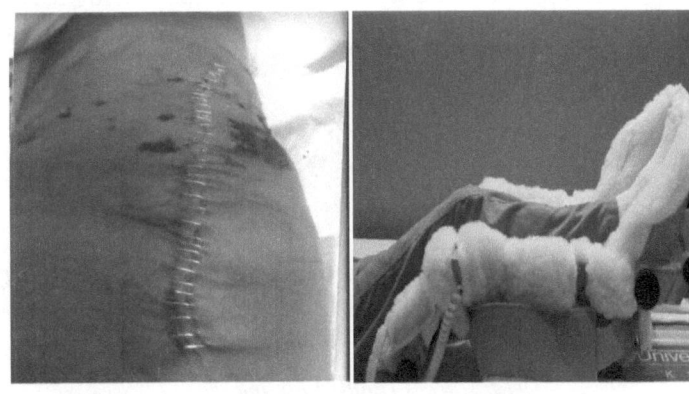

The Railroad leg   C.P.M. Machine

Since my leg was out of the CPM, I tried to shift myself in the bed. My leg was so heavy, it was difficult to move, but I was on a mission and determined to sit up on the side of the bed. First, I shifted my hips to the left of the bed, leaning against the rail. Gingerly, I moved the railroad track leg so it was close to my left leg. Once both legs were together, I shifted them to the left and held onto the bed rail for dear life. I was on the move.

Pain Medication

I was flying higher than the atmosphere can be seen with the naked eye. The combination of Oxy-Contin and Tylenol 3 really calmed my pain. However, it made me dizzy, drowsy, weak, and at the time it made me have crazy dreams and hallucinations. I quickly avoided the Tylenol 3. So I only took it when I was going to sleep. As time went by, I resorted to using less and less of both. Ibuprofen became my new best pain killer.

%%%

Nickel Mile (Reminiscent of Knee Replacement surgery)

I am sitting on this hospital bed,
with stitches running up and down my leg.
This self-imposed affliction
has been proposed for a while.
How could I have known that
I would be joining
a style I call the Nickel Mile.

The leg actually looks like a railroad track
that was never completed.
These adorable ice packs join my hospital bed,
with their cute little pouches neatly put
together with straps and strings. Oh,
here comes the pain, shooting up my leg
from both sides of the rails.

The thigh and leg used to be of normal size.
However, they now look like a giant walrus that's
relaxing on the beach; waiting to be crowned.
Like a balloon getting ready to float away,
my mind suddenly says "come...back...
balloon...I ...want...to...stay." I think
the medicine just kicked in.

The sheets and blankets lined up on the bed,
have invited a contraption that had to have
Come from outter space.
such a machine I never have seen.
Like a bridge coated with lamb's wool,
my leg is locked into
Place, so that this monster can gently
move the leg back and forth, back and
forth with such amazing grace.

The Nickel mile now has 25 little strips
that holds the railroad track in place.

Boy, isn't this modern medicine great?
I'm sitting here on my bed and the pain
has shot up to my head. The medication
must be administered again. Not fast
enough for me, but then again,
the slower it takes effect, the longer it
will take to have crazy dreams about all
the stylish people who have traveled
on the Nickel Mile.

©*February 1, 2017*

Surprise, Surprise, Surprise

As I was negotiating my own therapy of sorts, the therapist walked in and said,

"What a surprise. Look at you showing off…here, let me help you."

As she continued what I had started, finally, I was sitting on the side of the bed. I was quite pleased with myself.

The therapist sat a walker in front of me and smiled. I asked. "What's this for?" "You are going to stand for me and we're going for a walk!" she said.

I began fussing, "I can barely bend this knee, and you're talking about standing and walking…are you serious? I queried.

The therapist acted like she didn't hear me. She reached for my hands, placed them on the hand rail of the walker and said,

"Use the walker to support yourself. Slowly slide down the side of the bed to the floor. I want you to stand for just a minute...just to get used to it. Then I will guide you to taking your first steps."

The surprise was on me. Instantly, I got nervous...didn't think therapy would begin so soon. At least, I wasn't ready so soon.

The Walker

First, I held onto the rail of the walker...almost too tightly. I gripped it so hard, the therapist had to loosen my hands. As I held onto the walker for dear life, I slid my left leg (the good leg) down the side of the bed and down to the floor. I did the same to the railroad right leg. Once both legs were on the floor...

I winced and asked, "Can I get back in the bed now?"

Again, seeming to ignore me, The therapist said,

"Now, I want you to take a step with your left leg (the good leg). After which, you can take a step with your right leg."

As I followed her instructions, mentally in protest, I was shocked to find.... I could walk.

The therapist smiled and said, "You did good. Let's try to walk to the door."

"You want a lot, don't you?" I complained. These actions were done on the first day after surgery.

That walker took some time to get used to, but it became my best support mechanism. Therapy became a ritual for the next three days. I walked halls, uneven inclines, and even walked up and down stairways, as I would be doing at home. The therapist talked a lot about my range of motion and she measured how far back I could bend my knee. After in-patient and out-patient rehab, she said my range of motion should be at least 130 degrees.

Day Five Post-Surgery

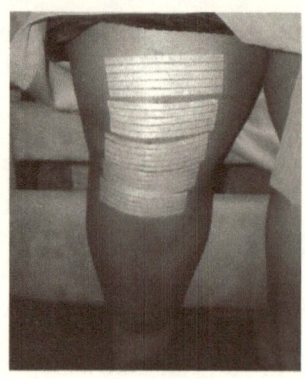

Since I knew, eventually, my left knee would have to done, when I saw the doctor, I asked,

"How will I know when to get the other knee done?"

He simply said, "Believe me, you'll know!"

On the fifth day of my stay at the hospital, I was released. My cousin Jean had agreed to come and take me home. When she came into my room, she could not believe that I was standing and walking with the walker.

She kept saying, "Girl you look good! You're doing so well."

Instead of us going right home, we stopped by a market to pick up food. I was hungry.

Home Away From Home

My husband had passed away four months before I had the surgery, so I was home alone. However, my youngest daughter Christina said,

"There is no way you can be by yourself. Why don't you come and stay with us?"

Us was she, her husband, son and daughter. They lived in an apartment. The only steps I would need to take were outside. I agreed and promised to behave and not be a bother. I soon became glad that I had agreed to stay with her. It benefited both of us.

She cooked my food, helped me with changing dressings, and made a fuss over me. When she needed to make quick errands, while the children napped, I freed her to do that, just by being there. By the time two weeks had passed, I was ready to go home, just because I wanted to sleep in my own bed.

A Night of Singing and Poetry

In previous months, I had been asked to perform my poetry at my church for an afternoon of singing, instrumental music, and spoken word. I wasn't even sure if I was going to make it to the presentation since I knew I was having the surgery.

A friend of mine had purchased a walker on wheels. Two week after the surgery, I walked into my church with the fancy red walker. It had a seat, and a basket for my goodies. When my church family saw me, they were shocked and couldn't believe I made it to the program.

I had to perform from the choir loft, so the deacons parked my walker and guided me by holding both my arms. They were concerned about the steps, but I maneuvered them like a pro. After the program was over and closing remarks were made, the chairman of the auxiliary spoke to the audience, saying that it was a blessing to have me there. They had changed the date, and she was concerned that I wouldn't be able to make it.

Then she said, "Isn't God great? Look what He can do."

Future Therapy

Initially, a therapist came to my home the first two weeks, and then I went into out-patient therapy where I had to go to a site near my home. The therapist who came to my home really worked with me; I liked her techniques. Once I began seeing a new therapist, I wasn't quite as pleased. Therapy is no joke, and it is hard work; I accept that some of the exercises and manipulations may be uncomfortable, but this darn woman was a sadist. She seemed to enjoy inflicting pain…especially on me. One time she was doing something with my knee cap that felt like she was trying to pull it out of my skin. I was in pain for three days. I was very upset with her. I contacted my doctor and asked him if I could start going to water aerobic. By this time, four weeks had passed and my skin had healed, so he said yes. I didn't mention my issue with the therapist.

My self-therapy, of going to water aerobics, was perfect. When my doctor saw me at eight weeks, he said,

"Your range of motion is fantastic." That was in 2008. Two years later, 2010, when my left knee could no longer support me, I had the surgery done with the same perfect results.

In both cases, I became a new woman. I could walk, do exercises with no pain, swim, and get around freely like I used to. I wasn't an 80-year old woman in a 56-year-old body anymore. My only prayer was that my knees last forever…even though I may only get 15 to 20 years out of them, if I take care and not abuse them. I will be happy to stay on the move as long as I can.

%%%

Medical Conference Call

As the head of this body, it has come to my attention that some changes have to be made. Many units will participate in the change and unfortunately, some of you will have to go. Now, these changes will be done in two increments. Please wait, until all information has been provided, before any comments are made.

We have found that some of you are not supporting the body and are causing shifting, swelling, and stiffness where there should be full cooperation from us all. Therefore, knees, you will be replaced with new, strong nickel and plastic based prosthetics that will support this body more effectively. The ankles and hips have been complaining for years; it is in our best interest, to make these changes.

Any comments…?

The hands exclaim that they didn't mind using canes and walkers to help out. But the left ankle angrily yells,

"The old knees almost caused us to drown in three feet of water, and I'm glad for the change."

Ok, having heard your views, the changes will go into effect immediately. The results of these changes means the body will no longer be in the one stepper club, the hands will no longer need canes or walkers, water aerobics will no longer be a problem and once again, the body will have full cooperation with itself. Therefore, this medical conference call is adjourned.

© 12/2008

This page is blank on purpose

VI. Pacemaker Society

Source: http://www.bostonscientific.com/en-
US/products/pacemakers.html

"Older people sit

Down and ask,

"What is it?"

The boy asks,

"What can I do with it?"

Steven Jobs

This page is blank on purpose

Pacemaker Society

The flu had its way with me for an entire week. However, I had come through it and had given myself several extra days before I resumed my regular activities…which were numerous. However, the one thing that was on my mind was the upcoming poetry and guitar concert at the local older adult place to be, the Center in the Park in Philadelphia, PA.

It was a Wednesday morning, the sun was so bright, like the middle of the summer; I just knew it was going to be a wonderful day…even though, my sister, Dar felt I should not go out anywhere.

She protested, "You should stay at home."

I was flying high on adrenaline and cabin fever.

My partner-in-rhyme and I were talking in the lobby of the center.

I said, "I feel weird."

She asked, "Weird… How?"

I simply said, "I don't know how to explain it, my equilibrium is off."

My partner felt that after eating lunch, I might feel better.

The lunch room looked like it always did...uninteresting. With its yellow walls and non-descript floors, tables and chairs, serving stations, orange water jugs, water pitchers on the tables, salt and pepper shakers and senior citizens anxious to receive their lunch. My food tray was placed in front of me. At first, I just looked at it. It didn't look appealing to me. The chicken parmesan, broccoli, brown rice, rolls and butter, and the cup of juice, looked back at me. I took one bite of the chicken parmesan and took it out of my mouth. All I tasted was salt. The broccoli, rice and rolls were lunch! My partner saw the expression on my face and knew I was neither happy nor satisfied. When we finished our meal, we headed to the auditorium where the concert was to be held. I began talking to the poets and letting them know what position they held on the program However, I still felt like crap, as if part of me was fading away.

To my partner I said, "Hey, if I don't feel any better, can you take over for me?" She agreed to help.

The time had come to begin the program. All of the performers were in place…the poets, the guitarist, photographers and most importantly…the audience. The podium and microphone were sitting on somewhat of an angle to the left of the auditorium. I stood up to go to the podium, but as I walked forward, I felt very weak.

At that point I prayed, "Lord, please, let me introduce these three people. Then, I will sit down."

As I introduced the poets, the first poet approached the podium. I stepped away to walk back to my seat, but instead, I began walking backwards. Just before I hit the wall that was behind me, the poet who was coming to the podium caught me and kept me from hitting the wall.

I was escorted to my chair, (from this point my story is hearsay), I was seated next to one of the poets, Ms. Ada. She said that I asked her if I could put my head on her shoulder. Months later, Ms. Ada recalled the event:

"Child, you scared me to death, because as soon as your head laid on my shoulder, your body went limp and you slid out of the chair onto the floor."

To this day, when she shares the story she says, "It's as if you died right there!"

Apparently, the nurse on staff at the Center in the Park was summoned. She could not get me to respond, though I had a pulse. 911 Emergency Medical Team (EMT) was called for assistance. The EMT did not fare any better than the nurse. I was transported to the local hospital. Upon arriving at the Einstein Medical Center's Emergency room, I understand that the doctors worked on me to no avail. They too could not get me to respond. It was as if I were alive but taking a long winter's nap.

I awoke suddenly in fight mode, to very bright lights, not knowing where I was. A man in a white coat was standing over me, but I didn't know who he was, or where I was. With fist in the air, I was swinging and this man began asking me questions.

"What's your name? Where were you born? When's your birthday? What year is it?

My response was non-committal. "Why are you asking me questions? Who are you and where am I?"

When the E.M.Ts left the Center, my partner-in-rhyme rode with them to the hospital. She had been sitting outside the room that I was in. Praying had been her solace for I had terrified her. She began thinking I was dying. When she heard me fussing at the doctor, she hollered.

"She's alive, she's alive!"

The doctor goes on to say, "Mrs. Peurifoy, you are in Einstein's Medical Center. You blacked out while hosting a show. We had a difficult time getting you to respond.

So, we sent for the electro-physiologist. I will let them know that you're awake, so they can come in and talk to you."

I retorted, "I want to go home; I am perfectly fine."

The doctor ignored me and left the room.

A new doctor entered the room. He was quite handsome, and he had somewhat of a Polynesian complexion. He smiled at me and said,

"Well, hello sleepy head. How are you feeling?"

Again, I retorted, "I'm fine. Now can I go home?"

"Mrs. Peurifoy...did I pronounce your name correctly?" He asked, "Yes. What's going on? Why can't I go home?" I asked.

Dr. Cutie, my name for him, went on to explain.

"Mrs. Peurifoy, you are in pretty bad shape. I understand, according to your daughters who has been notified that you are here, that you have blacked out before…at least three other times. Talk to me about that."

I mellowed in my response; "I don't know what happened. Each time, one minute before, I was fine; then the next… I blacked out for what I believed was no reason. At least that's how I felt because each time, I went to a hospital, they never had any answers."

It turned out that Dr. Cutie had the answer.

> He said, "Let me explain something to you. Our hearts have a generator which has four ventricle leads. You only have one working lead… that's trying to do all of the work for the missing leads. When it gets over worked, everything slows down…then you black out! It seems, my dear, that you need a *Pacemaker*."

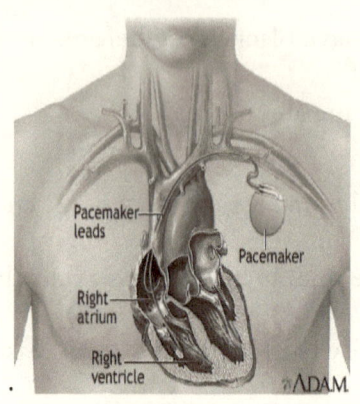

Source:

https://medlineplus.gov/ency/article/007369.htm

Of course, I was confused. But more importantly, I was upset that through all of the blackouts and hospitalizations, that none of the other doctors could figure that out.

I spouted, "Well, Doc, suppose I don't want a pacemaker?"

The doctor smiled at me and said, "Victoria, if this happens again, your heart stops, and you are no more."

Surprised, I said, "I'll take the pacemaker for $200, please."

Dr. Cutie laughed and said, "It will cost a little more than $200.00."

My Uncle Buck had a pacemaker and a defibrillator. I recall it looking like a transistor radio, about the size of a basic size smart phone of today, which protruded from his chest. The defibrillator would shock him if he got overly excited and the pacemaker paced the rhythm of his heart.

Of course, I had questions, "Doc, how big are the pacemakers now?" I asked (Now was 2010).

Dr. Cutie explained that the pacemaker wasn't much bigger than a half dollar and though it would be under my skin, in the region of my heart, it would not protrude. I wanted to know specifically what they would need to do and how the pacemaker worked.

He said, "We will implant the pacemaker and replace the three leads that burnt out. The function of the pacemaker is to help control abnormal heart rhythm. It will use electrical pulses to get your heart to beat normally; so that you won't be passing

out anymore."

Down

I'm wet.

How did I get wet?

I'm down.

How did I get down here?

Hand is bruised.

Hand is dirty.

How did I get down here...on this asphalt?

Vanity is out of the window

cause the doggone coat and pants

are a mess.

Oh boy, oh boy is this going to be

my claim to fame?

Can't stand up right,

walking seems to be a challenge.

Down I go again, just like once before in time.

Is it my heart, my head or my mind?

No one seems to know.

Down like water in a pipe I spiral, spiral down

onto a surface ...unsure... can't say why.

Against me, this ground has conspired.

Don't want this to occur no more.

Keep only my feet upon this ground

cause answers to my questions

seem to evade me

I'm down, I've falling down again.

Please someone, help me, help me,

oh God, please pick me up.

© 2010

Aftermath

In the hospital, everything went well. The pacemaker was working. They had a way of testing that was bizarre. The doctors placed these tags on my chest and monitored the pacemaker. Sometimes during this process, I felt like my heart was in my throat...but that was considered normal. I was released from the hospital with an appointment for a week from that day.

There was only one problem; after I left the hospital, I couldn't keep my food down. I was staying with my daughter Christina. She had insisted that I stay with her and her family, I did. I had to use a special pillow under my arm, on my left side, to cushion bumps or pain.

One night, my daughter had made a nice dinner for the family and I was ready for her vegetable strudel, fried tilapia, and broccoli along with dinner rolls. As I ate and watched TV, I was in a grateful mood. Half an hour later, I felt a lump in my throat and then a burning sensation in my chest. I got up and headed toward the kitchen to get a bag. However, as I passed the bathroom, something said…*make a right turn…NOW*. I landed on my knees, at the commode, relieving myself of my entire dinner…what a waste. I hurled everything to the point where I was experiencing dry heaves. Though I spoke to the doctor about this, I was told there was no correlation to the surgery. I disagreed. These events never happened before the pacemaker was implanted. For several months, when a tickle invaded my throat, it was always my warning that something was about to happen. I kept thinking, *is this the way that I'm going to have to live for the rest of my life?*

Once the heaving stopped, the incision was approved as healed.
I had to learn to remember that when I went to an airport I had
to say,

"I have a pacemaker."

Now, I can't walk through metal detectors anymore in life.

Eight Months Later

Instead of going back and forth to the hospital to have my
pacemaker checked, the doctor said, he believed I was ready for
remote examinations from home. I was sent a container, not
bigger than a shoe box. Inside the box were two wrist bands that
look like watches…except that instead of faces, they each had
buttons with a wire attached to it. In addition, there was one
round disk about the size of a tomato paste bottle top and about
a half an inch thick, but it looked like a donut. At first, I was
concerned about possibly getting shocked by electrical currents.

Pacemaker Monitor   **Monitor diagram**

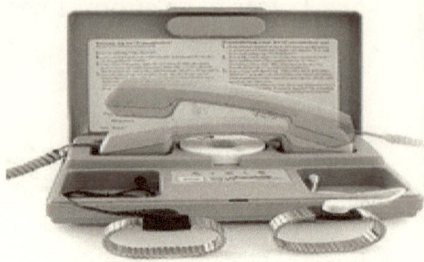 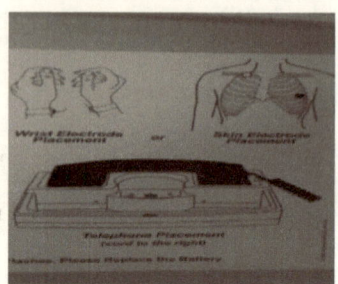

Source:

http://www.mednethealth.net/products_pacemaker.shtml

Then, there was the carriage that could accommodate the

receiver of a regular house phone. When the telephone is on the

carriage, various clicks and ticks would be registered to the

specialist who was on the other end of the phone giving me

instructions.

The specialist would say, "Take the disc and place it over

the pacemaker on your skin. Put the wrist bands on, and then

plug them into the box. After that, place the phone receiver

onto the carriage." I would sit there in anticipation of the results.

The technician said, "Okay, you can remove the bands from your wrist, put the disc back on the carriage and remove your phone receiver…everything is fine."
She has said this during every examination so far.
After I got my pacemaker, I discovered that many people I knew had pacemakers as well. They have had them for years. Two women even had new batteries put in after 10 years…and they are still alive both in their 90's and going strong.

When my grandchildren ask, "Grandmom, what is that thing on your chest (as they point to the scar on my chest)?"
I simply say, "This is the key to my heart."
I truly thank God for giving doctors the knowledge and skills. In addition, I'm glad that God is not through with me yet. Many years have passed since that fateful day. My pacemaker is used to me, and I am used to it. Finally no more dry heaves.

%%%

She's Alive

Who is that lady? She's in distress.
She has fainted and she slipped out
of a chair. She's in bad shape.
Fear is on everyone's face and in
their eyes and in their emotions.
heir looks all tell the story.
Gasps are heard throughout the room.
Will she
live, will she be alright?
All who know her, stressed through tears.
Her partner-in-rhyme is frantic.

Frightened by unknown surroundings,
an internal overdrive distress signal,
propelled her into heart palpitating anxiety.
She awakes suddenly. She goes into
fright, fight, and flight mode. The doctor
She questions...
who are you? She wonders...
Why is this stranger so close...?
Who is he...?
How did she get here...?
Why is she here...?

Where is here…?
Her Partner yells, "She's alive."

Sick, doesn't describe her situation.
As told to her… her illness was more like…dire.
A resolve awaits her answer. Frightened again,
but wanting to live, she says, I'll take the
Pacemaker for $200, please.
Candor still exists within her soul.
Living takes on a new meaning.
She prays, gives thanks, and says to herself,
God must not be through with me yet.

@ February 6, 2017

Two To Three Years

Two to three years the cysts attacked
Two to three years…then maybe you will get pregnant
Two to three years of having pains in the neck -
If no surgery - you won't be able to walk
Two to three years you will be blind
Two to three years, your knee won't serve you anymore.
Two to three years of black-outs those were no fun.
Two to three more years always sounds like trouble.

© March 15, 2017

Final Say

"I'm a grace case…God receives the glory."

Rev. Dr. Charles Goodman, of Tabernacle Baptist Church

I have had my say. I hope that you have learned, understood, and appreciated the patient and what they can go through before, during, and after any medical procedure. My story is also my testimony of strength and determination. I was blessed because through it all…I made it because of God's grace and mercy toward me.

Psalm 9:10 *Those who know your name trust in you, for you, LORD, have never forsaken those who seek you. Source: Holy Bible*

In each of these narratives I mention God, or how I had prayed to him for myself and for others, especially, the medical community. Trusting in God and believing on Him has been the only way I could ever made it: even though it is obvious that many times, I was scared to death of what my future held. Once I gave up the pity party and left the burden with God…it was a wrap. He took care of everything.

Okay, I will get off my soap box, but I had to give God the Glory, honor, and praise for all he has done for me and my family. God bless you.

Who are the folks quoted at the beginning of each narrative?

Mahatma Gandhi – Anti-War Leader

Was born in Par Bandar, India. He was an Indian lawyer, politician, social activist, and writer who became the leader of the Nationalist movement against British rule of India
Source: https://en.wikipedia.org/wiki/Mahatma_Gandhi

Martin Luther King – Preacher

Martin Luther King Jr. was an American Baptist minister and activist who was a leader in the Civil Rights Movement. He is best known for his role in the advancement of civil rights using nonviolent civil disobedience based on his Christian beliefs.
Source: https://en.wikipedia.org/wiki/Martin_Luther_King_Jr.

Paul "Steve" Jobs

was an American entrepreneur, businessman, inventor, and industrial designer. He was best known as the co-founder, chairman, and chief executive officers of Apple, Inc. and pioneer of the personal computer revolution.

Source: http://www.imdb.com/name/nm0423418/bio

Muhammed Ali – Heavy Weight Boxer

Muhammad Ali was an American professional boxer and activist. He is widely regarded as one of the most significant and celebrated sports figures of the 20th century.
Source: https://en.wikipedia.org/wiki/Muhammad_Ali

Maya Angelou – Poet

An American poet, memoirist, civil rights activist. She published seven autobiographies, three book of essays, several books of poetry and was credited with a list of plays and movies.
Source: https://en.wikipedia.org/wiki/Maya_Angelou

Victor Hugo – French Poet

Was born 1802 in Besan'con, France and died in Paris France 1885. He trained to be a lawyer; before he embarked on a literary career. He was a poet, novelist, and dramatist, also wrote French romantic books. He is best known for the novels, Les Miserables (1862) and Notre-Dame de Paris (1831).
Source: https://en.wikipedia.org/wiki/Victor_Hugo

Glossary

Merriam-Webster Medical Vocabulary

(Unless otherwise mentioned definition sources are from the
Merriam-Webster Medical Dictionary)

Aspirate - as·pi·rate:
to remove (a fluid) from a body cavity by use of an aspirator orsu
ction syringe.

Basal temperature: the temperature of the body at rest that is
typically taken immediately after waking from sleep

NOTE: The basal body temperature of a woman slightly rises
during ovulation and may be tracked to estimate when future
ovulation will occur.

When she ovulates, her basal body temperature rises about one
degree and remains elevated until her next period.—Jane E.
Brody, The New York Times, 1 Jan. 2002

Benign: *of a mild type or character that does not threaten health or
Life* benign malaria benign liver cyst; especially: *not becoming
cancerous* a benign lung tumor—*compare malignant* 1

Biopsy: the removal and examination of tissue, cells, or
fluids from the living body

Cat Scan: A computerized tomography (**CT**) scan combines a
series of X-ray images taken from different angles and uses
computer processing to create cross-sectional images, or slices,
of the bones, blood vessels and soft tissues inside your body. **CT
scan** images provide more detailed information than plain X-rays
do. a picture of the inside of a part of your body that is made by
a computerized machine

Cerebrospinal Fluid - ce·re·bro·spi·nal flu·id: a colorless liquid that is comparable to serum, is secreted from the blood into the lateral ventricles of the brain, and serves chiefly to maintain uniform pressure within the brain and spinal cord

Continual Passive Motion Machine... aka CPM: is used after various types of reconstructive joint surgeries such as knee replacement and ACL reconstruction. Using a CPM machines can speed recovery times by allowing for better diffusion of nutrients into damaged cartilage as well as diffusion of other products out. CPM also prevents build up of scar tissue, which when present, can decrease a joint's range of motion.

Cyst: a closed sac having a distinct membrane and developing abnormally in a body cavity or structure

Dilation and curettage aka D&C: A medical procedure in which the uterine cervix is dilated and a curette is inserted into the uterus to scrape away the endometrium (as for the diagnosis or treatment of abnormal bleeding or for surgical abortion during the early part of the second trimester of pregnancy)—called also D&C

Defibrillator - de·fib·ril·la·tor: an electronic device that applies an electric shock to restore the rhythm of a fibrillating heart.

Dexamethasone - dex·a·meth·a·sone: is used for Ovulation Induction. Anovulation is one of the more common causes of **infertility**. Clomiphene citrate (Clomid) is a medication that helps with ovulation, but only about 80% of women respond. It is a medicine that makes Clomid more effective for ovulation.

Endoscope: an illuminated usually fiber-optic flexible or rigid tubular instrument for visualizing the interior of a hollow organ or part (as the bladder or esophagus) for diagnostic or therapeutic purposes that typically has one or more channels to enable passage of instruments (as forceps or scissors)

Fallopian Tubes: either of the pair of tubes that carry the eggs from the ovary to the uterus—called also uterine tube

Field of Vision Test (F.O.V.): The visual field test is a subjective measure of central and peripheral vision, or "side vision," and is used by your doctor to diagnose, determine the severity of, and monitor your glaucoma. The most common visual field test uses a light spot that is repeatedly presented in different areas of your peripheral vision.

Fibroadenomas : These are the most common benign tumors. If you push on them they are solid, round, rubbery lumps that move freely. They're usually painless. Fibroadenomas happen when your body forms extra milk-making glands. Women between 20 and 30 get them most often. They're also more common in African-American women.
Source: http://www.webmd.com/breast-cancer/benign-breast-lumps#1

Flexuril: (Cyclobenzaprine): is a muscle relaxant. It works by blocking nerve impulses (or pain sensations) that are sent to your brain. It used together with rest and physical therapy to treat skeletal muscle conditions such as pain or injury.

Source: https://www.drugs.com/flexeril.html

Frontal lobe: The part of each hemisphere of the brain located behind the forehead that serves to regulate and mediate the higher intellectual functions. The frontal lobes are important for controlling thoughts, reasoning, and behaviors. Source: http://www.medicinenet.com/script/main/art.asp?articlekey=25285

FSH: follicle-stimulating hormone: - a hormone from an anterior lobe of the pituitary gland that stimulates the growth of the ovum-containing follicles in the ovary and that activates sperm-forming cells—called also follitropin

Laparoscopy - lap·a·ros·co·py: 1. Visual examination of the inside of the abdomen by means of a laparoscope—called also peritoneoscopy

2. An operation (as tubal ligation or gallbladder removal) involving laparoscopy

LH: luteinizing hormone - a hormone secreted by the anterior lobe of the pituitary gland that in the female stimulates ovulation and the development of corpora lutea and in the male the development of interstitial tissue in the testis

MAAGNETIC RESONANCE IMAGING **MRI:** the procedure in which magnetic resonance imaging is used
Neurologist: one specializing in neurology; *especially*: a physician skilled in the diagnosis and treatment of disease of the nervous system

Occupational Therapy:
a form of therapy in which patients are encouraged to engage inv ocational tasks or expressive activities, as art or dance, usually in asocial setting.

Optic Chiasm – Optic kahy-az-muh:
A flattened quadrangular body that is the point of crossing of the fibers ofthe optic nerves. Also called *optic decussation*.

Oxy-Contin: (oxycodone) is a narcotic pain reliever used to treat moderate to severe pain. Includes **OxyContin** side effects, interactions and indications. Source:
https://www.drugs.com/search.php?searchterm=Oxy-Contin+

Pacemaker:
An electronic device implanted beneath the skin forproviding a n ormal heartbeat by electrical stimulation of the heartmuscle, used in certain heart conditions.

Physical Therapist: the treatment or management of physical disability, malfunction, or pain by exercise, massage, hydrotherapy, etc., without the use of medicines, surgery, or radiation.

Pituitary Tumor: Tumors arising from the pituitary gland itself are called adenomas or carcinomas. Pituitary adenomas are benign, slow growing masses that represent about 10% of the primary brain tumors. Pituitary carcinoma is the rare malignant form of pituitary adenoma

Polycythemia Vera - pol-e-sy-THEE-me-uh VEER: uh: is a slow-growing blood cancer in which your bone marrow makes too many red blood cells. ... They also cause complications, such as blood clots, which can lead to a heart attack or stroke. Polycythemia Vera isn't common. Feb 8, 2017

Prednisone: Prednisone is a corticosteroid. It prevents the release of substances in the body that cause inflammation. It also suppresses the immune system.

Prednisone is used as an anti-inflammatory or an immunosuppressant medication. Prednisone treats many different conditions such as allergic disorders, skin conditions, ulcerative colitis, arthritis, lupus, psoriasis, or breathing disorders.
Source: https://www.drugs.com/prednisone.html

Steroid: any of a large group of fat-soluble organic compounds, as the sterols, bile acids, and sex hor mones, most of which have specific physiological action.

Tylenol 3: Acetaminophen and codeine is a combination medicine used to relieve moderate to severe pain.

Acetaminophen and codeine may also be used for other purposes not listed in this medication guide.

Tubal Lavage: A flexible catheter originally designed for hysterosalpingography was used for intraoperative tubal lavage in 15 patients undergoing various infertility procedures. The catheter was easily inserted without the use of a tenaculum or cervical dilators, provided reliable dye instillation, and offered an alternative to the pediatric Foley or Buxton clamp.

Source: https://www.ncbi.nlm.nih.gov/pubmed/3336561

Ultra sound:

1: vibrations of the same physical nature as sound but with frequencies above the range of human hearing

2: the diagnostic or therapeutic use of ultrasound and especially a noninvasive technique involving the formation of a two-dimensional image used for the examination and measurement of internal body structures and the detection of bodily abnormalities —called also sonographer

3: a diagnostic examination using ultrasound

Urofolliyrophin aka uFSH - : a follicle-stimulating hormone

Vertebrae: any of the bones or segments composing the spinal column, consisting typically of a cylindrical body and an arch with various processes, and forming a foramen, or opening, through which the spinal cord passes.

Anthologies to which the author has contributed:

-Drexel – Dornsife – 2015-16 – Anthology II –Writer's Room
Issue 2
-Drexel – Dornsife – 2016-17 –Anthology III - Writer's
Room Issue 3
-Seniors' Rockin' the Pen – 2014 - Poetry and Discussion –

-Moonstone 2014 - 2017Anthology of Featured Poets –
Moonstone Poetry Series

-Poe-Trees 2014 - The Collective Poetry Group

New Poems within these narratives

Fear of Cancer
When?
Nurse Angeline
I can't see
25 years
Nickel Mile (formerly entitled the Titanium Mile)
She's alive

Poems in this book previously published

Medical Conference Call
Down

Other Books by the Author
Publisher: Lulu.com

Truth with Purpose is a poetry book written by two women who share the truth about life and its circumstances; as well as the purpose or sometimes, the consequences of our decisions...in poetic form. Courageous and honesty are evident on every page, but beyond that, this book stands as proof positive of compassion and a shared desire to make a lasting contribution to this world. These two earthbound muses have crafted the poetry within these pages with ease and dedication. Together they have shared verses poetically and unapologetically. This will be a book to place on your coffee table and let your friends read but they will want their own

Truth with Purpose: Paperback, ISBN 978-1-312-69665-5 **Poetry**

This children's book is educational and family oriented. It allows the parent to have fun with his child or grandchild, as they learn about the science of batteries at the child's level. It is about a little boy who has a personal battery man to supply the batteries he needs for his toys. After reading this book your child will feel confident about understanding how batteries work and how to properly use them in their toys that need the energy from **The Battery Man:** Children's Book, ages 3-5, ISBN 978-1-312-48178-7

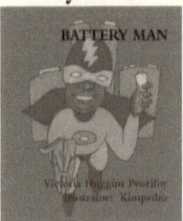

Children's Storybook

A BLADE OF GRACE
An Autobiography

VICTORIA HUGGINS PEURIFOY

In life we learn lessons in one of three ways: by experiencing them, observing them, or through the wisdom gifted to us by a loving elder. In Victoria Huggins Peurifoy's prose and poetry we are privileged to actually learn by all three methods. As a writer, Victoria (or as she is affectionately called "The Axiom") delivers her personal story so honestly and intimately you at times believe she is talking about you or you start to think you are her! But what ultimately comes through is the fact that she, as an elder, imparts the insights she has gained. Too often in today's literary world, memoirs or autobiographies become nothing more than sensationalized "tell all" exposés. Their only value comes from shock. Not true with "A Blade of Grace." Victoria tells her truth without making herself a victim, villain or vixen; which ultimately makes her a heroine. I have nothing further to say but; put on a pot of coffee, sit in your favorite chair, and enjoy! Elijah Pringle, Editor and Host of Poetry Quid Pro Quo

A Blade of Grace, ISBN 978-1-304-07614-4 (Hardcover) **Autobiography**

Poetry that considers what if running between the raindrops could wash away blown up pride. The author visits intimates observations such as Grandmom's Hands and what they mean. Thinking about one's heart and pain wreaks havoc. She lets everything around her generate poems that make you think...If the ocean could talk...what might it say. A female prisoner's Christmas in jail. Memoirs are added just for fun. If you like all types of poetry you will love this

Run between the Raindrops, Paperback, ISBN 978-1-312-71922-4 **Poetry**

This is a collection of inspirational poetry, prose and photography from the author. The photographer indicated is the Axiom, which is the authors "handle". We often question God, especially when things happen to us of which we are not satisfied. The author shares her outlook about having a religious imprint and how we can accept what God does for us. She speaks of how God is here for us, but we still don't know him. She attempts to give another way to think about praying such as Praying for nothing. She shares her love of Jesus, the Father and the Holy Spirit. Be inspired.

God's Calling, Paperback, 70 pages, ISBN: 978-1-300-33442-2 – Poetry

This is a story of a 92 year old woman who had some interesting stories to tell about childhood, marriage, children, history and traveling to and living in Japan. As her daughter indicated, talking about herself was out of her comfort zone, but our every Thursday afternoon sessions allowed her freedom to reveal her heart, love, life and the challenges she endured.

I Have Not Lived In Vain, Paperback, ISBN: 9781105798207 –**Autobiography** of 92 year old

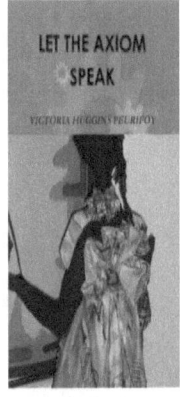

This book of poetry, stories, prose, photographs and love of life and truth are expressed as the author speaks of living an inspired life, how love may not be love, and how in life...stuff happens among the many headings. Laugh with her as she exposes her comedic side to how she looks at the situations that generate the poems

Let the Axiom Speak, Paperback, 201 pages, ISBN: 9781105570896 – **Poetry**

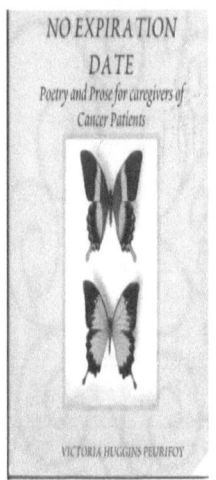

This is a book of poetry for people who are caregivers of terminally ill cancer patients. The caregiver is the last person that folks think about when it concerns a cancer patient; and they are the one who should have just as much support mentally, emotionally, spiritually and comforting. They try hard to do everything in their power to help and maybe in their mind even keep their loved one alive. It is a horrific decease, and prayerfully these poems will help the caregiver know they are not alone.

No Expiration Date, Paperback, 56 pages, ISBN: 9780557030477 **Poetry**

IN THE SPRING OF MY REMEMBRANCE is a book of prose, poetry, memoirs, and photography. The poet has chosen to break the poems down into categories that fit her moods such as "Inspiration, Love, Fun, Memorials, History/Political, these are just some of the categories. Photography is also a passion of the author and she shares a few photographs just for fun. This book includes pictures that were favorites, but in no way exhibits the many pictures that she has taken around her hometown Philadelphia, PA. Memoirs included in this book reveal another side of the author who believes that everyone goes through some things in their life that may resemble other folk's lives but talking about it or thinking about those situations just hurts too much. This is a small world and we all can relate to many times of happiness, experiments, experiences and tragedies that make up who we are. Victoria wants you to be inspired, have fun, have total recall and just enjoy.

-In the Spring of My Remembrance, Paperback, 153 pages, ISBN: 978-0-557-02740-8 - **Poetry**

"The Wild Spirit Behind the Voice", is a book of poetry and thoughts. These poems were written over a period of 30 - 40 years. Some of the poems are very simple, some are deep, some make you think, some are fussy, and some are just fun.

-The Wild Spirit behind the Voice, Paperback, 50 pages, ISBN: 9780557079278 – **Poetry**

About the Author

Victoria Huggins Peurifoy, is a Christian, who is an author, Poet, Spoken word artist, Storyteller, Ghost Writer, Photographer, Narrator, Consultant, Facilitator, Voice Talent, and Public Speaker.

Victoria is a graduate of Community College of Philadelphia. She has an Associate's Degree in Liberal arts. She studied Business, Accounting and Creative Writing. She retired from the Department of the Treasury, after 35 years. She is a twenty 29 year resident of the Germantown area of Philadelphia and is fondly called "The Axiom" for she speaks the truth. She is a self published author with twelve books, three chapbooks and three CD's to her credit. She is the Facilitator for the Poetry and Discussion and former Facilitator for the Best Day of My Life (So Far) a story writing group at the Center in the Park's older adult center; located in Germantown. Victoria also writes biographies for senior citizens. She assisted senior citizen poets put together an anthology entitled *"Seniors Rockin' the Pen.* She and her poetic partner in rhyme, Runett Nia Ebo *Co-host* Poetify Poetry to Edify, a family friendly poetry venue.

This author, has performed for Community College's Poetry event for 2017's Black History Month, The Big Blue Marble Bookstore, The Kroc Gems of the Salvation Army Kroc Community Center, Moonstone Poetry of Philadelphia, Black Expo at the Convention Center, The Philadelphia Jazz Project, The Clef Club, The Mt Airy Arts Garage, the Black Writer's Museum's Poetry Marathon and their Story Telling Saturdays, Moonstone's Poetry Quid Pro Quo, The Ethical Society, Germantown Jazz &Poetry Festival in Vernon Park, Pecolia Breedlove Tuesdays, and at the Freedom Theatre, just to name a few.

Victoria is a contributing author/poet to: Philadelphia Moonstone Poetry's 2015 thru 2017's anthology and Drexel University's Writer's Room 2015 to 2017 Anthology. In addition, she was also added to the "Poe-Trees" Anthology sponsored by "The Collective," a poetry group. She is a member of the Germantown Roundtable and contributor to their various poetry events. Victoria sings on the Church Chapel choir at her church, she is one of the voices of the Announcement Clerks at her church, and she is an active member of the Good Sheppard Circle. She is a widow with four children and eight grand children.

ISBN 978-1-387-09806-4   Paper Back